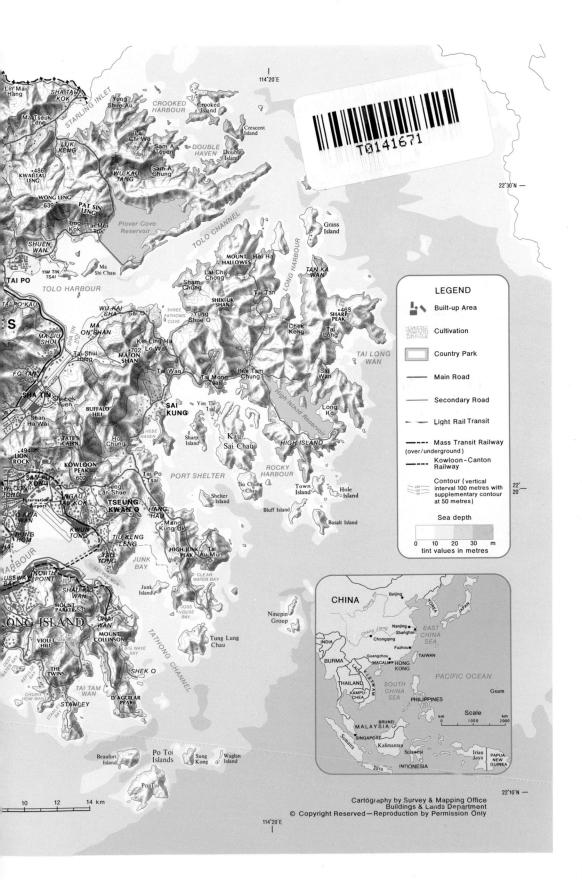

T0141671

LEGEND

- Built-up Area
- Cultivation
- Country Park
- Main Road
- Secondary Road
- Light Rail Transit
- Mass Transit Railway (over/underground)
- Kowloon-Canton Railway
- Contour (vertical interval 100 metres with supplementary contour at 50 metres)

Sea depth

0 10 20 30 m
tint values in metres

Lin Ma Hang
Sha Tau Kok
Ma Tseuk Leng
Yung Shue Au
CROOKED HARBOUR
Crooked Island
Crescent Island
STARLING INLET
LUK KENG
Lai Chi Wo
Sam A Tsuen
DOUBLE HAVEN
Double Island
Sam A Chung
•486 KWAI TAU LENG
WU KAU TANG
WONG LENG 639
PAT SIN LENG
Ting Kok
Tai Mei Tuk
Plover Cove Reservoir
TOLO CHANNEL
Grass Island
SHUEN WAN
YIM TIN TSAI
Ma Shi Chau
MOUNT HALLOWES
Hoi Ha
TAI PO
TOLO HARBOUR
Lai Chi Chong
Sham Chung
TAN KA WAN
TAI PO KAU
WU KAI SHA
Sai O
THREE FATHOMS COVE
Yung Shue O
SHEK UK SHAN
Tai Tan
LONG HARBOUR
•465 SHARP PEAK
MA LIU SHUI
MA ON SHAN
Kei Ling Ha
Lo Wai
Chek Keng
Tai Long
FO TAN
Tai Shui Hang
•702 MA ON SHAN
Pak Tam Chung
TAI LONG WAN
SHA TIN
Tai Wan
Tai Mong Tsai
Sai Wan
Siu Lek Yuen
BUFFALO HILL
Yim Tin Tsai
High Island Reservoir
Shan Ha Wai
SAI KUNG
TATE'S CAIRN
HEBE HAVEN
Ho Chung
Sharp Island
Kau Sai Chau
HIGH ISLAND
Long Ke
•494 LION ROCK
KOWLOON PEAK 602
Tai Po Tsai
PORT SHELTER
ROCKY HARBOUR
SAN PO KONG
Tseng Lan Shue
KWLOON TONG
NGAU TAU KOK
TSEUNG KWAN O
HANG HAU
Shelter Island
Tiu Chung Chau
Town Island
Hole Island
KWUN TONG
TIU KENG LENG
Mang Kung Uk
Bluff Island
YAU TONG
LEI YUE MUN
JUNK BAY
HIGH JUNK PEAK
Tai Au Mun
Basalt Island
CAUSEWAY BAY
NORTH POINT
SHAU KEI WAN
CLEAR WATER BAY
Junk Island
CHUNG HOM
MOUNT PARKER 531
CHAI WAN
JOSS HOUSE BAY
HONG KONG ISLAND
VIOLET HILL
MOUNT COLLINSON
BIG WAVE BAY
Tung Lung Chau
TATHONG CHANNEL
THE TWINS
SHEK O
TAI TAM WAN
D'AGUILAR PEAK
STANLEY
REPULSE BAY
DEEP WATER BAY
CHUNG HOM WAN
Beaufort Island
Po Toi Islands
Sung Kong
Waglan Island
Ninepin Group
Po Toi

---

10 12 14 km

114°20'E

22°30'N —

22°20' —

22°10'N —

CHINA inset

CHINA
Beijing
KOREA
JAPAN
Huang He
Nanjing
Shanghai
EAST CHINA SEA
Chongqing
Chang Jiang
INDIA
Fuzhou
TAIWAN
BURMA
Guangzhou
MACAU
HONG KONG
PACIFIC OCEAN
LAOS
VIET NAM
THAILAND
SOUTH CHINA SEA
PHILIPPINES
Guam
KAMPU CHEA
BRUNEI
MALAYSIA
Sumatra
SINGAPORE
Kalimantan
Sulawesi
Irian Jaya
PAPUA NEW GUINEA
INDONESIA
Java

Scale
km 0 1000 2000 km

Hills and Streams
An Ecology of Hong Kong

Hills and Streams

An Ecology of Hong Kong

David Dudgeon
and
Richard Corlett

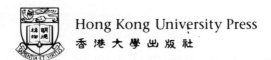

Hong Kong University Press
香港大學出版社

Hong Kong University Press
139 Pokfulam Road, Hong Kong

© Hong Kong University Press 1994

ISBN 962 209 357 4

Printed in Hong Kong by Prosperous Printing Co., Ltd.

This book is dedicated
to our wives,
May and Tim.

Contents

Contents

Preface

This book is about the ecology of freshwater and terrestrial habitats in Hong Kong. Those wishing to know something of the streams, woodlands, shrublands and grasslands have had few sources to turn to, apart from G.A.C. Herklots' account, *The Hong Kong Countryside* (1951), and S.L. Thrower's handbook, *Hong Kong Country Parks* (1984), both now out of print. Our main aim was to write a book that would go some way towards filling this information gap. We have deliberately omitted the largely urban or agricultural coastal flatlands, because the ecology of these highly-managed areas is fundamentally different from the semi-natural uplands that we cover. Even here, however, we make no claims about completeness of coverage. Inclusion of topics has depended upon our own interests, and the research undertaken by our colleagues and graduate students at The University of Hong Kong. Much has yet to be learned about Hong Kong's terrestrial and freshwater habitats. If this book stimulates further investigations, then our efforts will have been amply rewarded.

A second reason for writing this book was to serve the needs of local conservation. In common with many other people, we are alarmed by the accelerating despoliation and destruction of the Hong Kong countryside. Mere documentation of Hong Kong's ecology will not, on its own, prevent habitat destruction, but ecological knowledge is the basis of sound conservation planning and management. Only if we can inspire others towards conservation action and enhance their knowledge and appreciation of the countryside, will it be possible to preserve some of our heritage.

As we are university teachers, we have written primarily with undergraduates in mind, but the text is not aimed only at those specializing in ecology. School teachers and naturalists should also find this book useful and, we hope, enjoyable. Inevitably, the text is a compromise between providing sufficient depth and detail to interest the trained biologist, and enough basic material to aid the non-specialist. Our success in this regard must be judged by the reader. Despite the fact that some readers may see them as intimidating, we have been unable to avoid the use of scientific names (latin binomials) for organisms. Where there are widely-used names in English, we have adopted them, but have also given the scientific name the first time that a particular plant or animal is mentioned. For the great majority of Hong Kong organisms, however, there is either no English name at all or the scientific name is more widely known.

We emphasize that this is not an ecology textbook and we have not attempted to cover all facets of modern ecology. For that, the reader must look elsewhere. Ecological concepts and theories are introduced only where they are necessary to understand the features of Hong Kong's ecology that we discuss. A glossary has been added at the end of the text to define those technical terms that we have been unable to omit.

While we have tried to avoid arid lists of facts and species names, or the presentation of ecological concepts for their own sake, the theory of evolution by natural selection — the single most important generalization in biology — is employed liberally to provide context and explanation for the phenomena which are described in these pages. The geneticist T.H. Dobzhansky wrote that 'Nothing in evolution makes sense except in the light of ecology'; equally, nothing in ecology makes sense except in the light of evolution. For this reason, we have devoted a few pages of this book to an account of the theory of evolution by natural selection. While there are no surprises for biologists in these pages, we hope that readers unfamiliar with the theory will find the material a useful aid for interpreting ecological patterns and processes in Hong Kong. 'The trouble with biology', evolutionist John Maynard Smith is reputed to have said, 'is that there are too many goddamn facts!' Evolution provides a structure within which these facts become understandable and, to some extent, predictable; that is its importance for ecologists and natural historians.

In sum, this book considers *what* patterns appear when the terrestrial and freshwater habitats of Hong Kong are studied, and discusses *why* these patterns arise. The answers to the second question depend upon a combination of influences: the evolved features of organisms, present-day interactions between organisms and their environments, and the historical changes that have occurred in Hong Kong over the centuries. Ultimately, we hope to make the ecology of Hong Kong's hills and streams more understandable, and so further the reader's appreciation of these environments.

Acknowledgements

This book could not have been written without the assistance, advice and information provided by our students (past and present) and colleagues at The University of Hong Kong, as well the many amateur naturalists who shared their unpublished data with us and made helpful, and sometimes inadvertent, suggestions. In particular, we would like to acknowledge Gary Ades, Gloria Barretto, Dr Mike Bascombe, Andrew Benton, Fr. Anthony Bogadek, David Carthy, Anjali Chandrasekar-Rao, Lawrence Chau, Chong Dee-hwa, John Fellowes, Tony Galsworthy, Nick Goodyer, Dr Ron Hill, Ho Ching-yee, Ho Wai Hoong, Dr Richard Irving, Amy Lau, Michael Lau, Bill Meacham, David Melville, Dr Mervyn Peart, Graham Reels, Winnie Tang, Dr David Workman, Richard Webb, Keith Wilson, Lew Young and Michelle Zhuang. Billy Hau of WWF(HK) supplied indispensable advice on conservation and land-use planning legislation.

Gary Ades, Carmen Anderson, Andrew Benton, Lawrence Chau, John Fellowes, Michael Lau, Graham Reels, Lew Young and (especially) David Melville made many useful suggestions based on their reading of a draft manuscript, thereby contributing to the intelligibility of the final text. We are grateful to them, and to Gary Ades, Chong Dee-hwa, John Fellowes and Michael Lau for permission to use their photographs in this volume.

1

Evolution and Adaptation

In order to understand why the plants and animals in Hong Kong live where they do, and why they have adopted particular habits and life-styles, it is helpful to have some knowledge of how evolution has shaped them. Thus this chapter is devoted to a brief account of evolution by natural selection and its importance as a principle underlying ecology. The latter point is illustrated clearly by the number of times that the words 'evolution' and 'evolutionary' are used in the following chapters.

Because the evolutionary drama has been played against the backdrop of Hong Kong's particular environment and history, this prologue is followed by a chapter which describes the climate, soils and geology of the Territory [i.e., the Territory of Hong Kong] and gives an account of the environmental changes brought about by human activities over the centuries. In essence, the interactions between evolutionary processes, human activities, and the particular physical and climatic features of Hong Kong — acting in consort upon the local flora and fauna — have shaped the ecology of hills and streams described in Chapters 3 to 8, and have given rise to the need for conservation that is highlighted in Chapter 9.

Origins

The earth is at least 4.5 billion years old. Life arose over 3.5 billion years ago but, for much of its history, the principal players were prokaryotic organisms with cells which, unlike our own, lack a cell nucleus and certain other structures typical of 'higher' organisms. It was not until about 1.4 billion years ago that eukaryotic cells of the type making up our own bodies appeared, and the oldest multicellular fossils are no more than 700 million years old. These fossils are unlike any organisms living now. It is only around 550 million years ago that groups of animals whose descendants are alive today appear in the fossil record. These modern life forms occupy only about 15% of the total time that there has been life on earth. We, of course, have an even shorter history: if the earth's geological chronology could be conceived as lasting one year, then our own species — *Homo sapiens* — would have been present for less than the last second of that year. Humankind, and the diverse organisms that inhabit environments in Hong Kong and elsewhere, are products of evolution. An understanding of this process is essential for anyone who wants to understand nature.

Natural selection

Evolution is the unifying principle underlying ecology and, indeed, all biological science. All organisms, whether they live on land or in water, are products of evolution. How does this happen? Charles Darwin's theory of natural selection provides the only plausible mechanism which can account for biological evolution. Darwin realized that, as a consequence of natural variation, some individuals will be more likely than others to obtain food and other resources, and so to survive and breed in a given environment. These differences have the consequence that individuals most suited to prevailing conditions will leave more offspring than those which are less suited, so contributing more descendants to succeeding generations. Some of these differences will be inherited by the offspring of successful individuals, which will, in turn, leave many descendants resembling them. Over a number of generations, there will be a cumulative change as the characteristics of successful breeders spread through the population; natural selection of the fittest will have taken place. The characteristics of individuals comprising a population will thus alter or evolve from generation to generation, with the result that they tend to become more adapted to local conditions.

In a sense, the environment acts as a filter: only those individuals with appropriate morphologies, physiologies and behaviours can survive and breed. They leave descendants which are suited to conditions that their parents experienced and which share some of the characteristics (adaptations) which made their ancestors successful. The term adaptation refers to the possession of features of form, behaviour and bodily function which contribute to success in terms of survival and reproduction. In other words, an adaptation is a 'good thing' or an appropriate design feature — the sort of characteristic that an intelligent creator might have provided.

Regardless of the nature of the adaptations evolved, because of the environmental 'filtering' effect we expect to see a match between organisms and local conditions in an environment. For example, we would anticipate that the inhabitants of a slow-flowing lowland stream would differ from those of a mountain torrent (indeed such observations have prompted schemes of river zonation and classification; see Chapter 6: Longitudinal zonation), and we would expect denizens of two such torrents in separate mountain ranges to possess similar adaptations. It is for this reason that animals in stony upland streams the world over possess comparable (but not identical) adaptations. Note, however, that adaptive change lags behind environmental change; hence adaptations do not evolve to meet the present needs of organisms, but arise from the success of ancestors that pass through the environmental filter. Strictly then, organisms are not *adapted to* (Latin *ad* = to) their current environment, but *abapted from* (Latin *ab* = from) ancestral environments.

The fact that adaptive change lags behind environmental change ensures that evolution by natural selection cannot be a perfecting process. The physical characteristics of environments change continually (as exemplified by the Ice Ages and global warming), and organisms must evolve to suit new conditions or they will perish. Even if the physical environment remains constant over time, the biological environment changes because the plants and animals that are the food, competitors or natural enemies of any species are themselves evolving. The process is complicated further by the fact that migrations or invasions by predators and competitors will alter the selective pressures experienced by a particular species. In this context, it should be borne in mind that most native Hong Kong species are presumably abapted from the primeval forest which once covered the Territory, and not adapted to modern, man-modified environments.

Conflicting demands

Even if all aspects of the environment remain constant, simultaneous perfection of every adaptation of an organism by natural selection is impossible because resources are limited. For example, the number of eggs that an insect such as a dragonfly produces cannot be maximized at the same time that egg size (investment per egg) is maximized. The development of a particular adaptation should not be seen in isolation, but in the context of the organism as an integrated whole. Maximizing egg production in a dragonfly might be incompatible with adult flight ability, and aerodynamics will be traded-off against fecundity. Many attributes will influence an organism's success at producing viable offspring, and it will never be possible to maximize all fitness components simultaneously. There will be trade-offs and constraints so that the fittest individuals will optimize, rather than maximize, fitness components. Thus, the appropriate trade-off between number and size of eggs will vary under different ecological circumstances, and the fittest individuals will allocate energy and other resources in a way that achieves the optimal balance between these two parameters.

Ecologists have adopted optimality theory from economics and engineering as an aid to understanding adaptation. The application of optimality theory depends on a knowledge of the costs and benefits of possible adaptations in particular ecological circumstances; we assume that natural selection has favoured the adaptation which maximizes the ratio of benefits to costs. For an insect living in an ephemeral or short-lived habitat, such as a water beetle in a rain pool, the optimum balance between investment in eggs and in flight ability will be in the direction of increased dispersal. The cost of the extra weight of eggs will not be repaid in increased fitness if the insect cannot fly far enough to colonize a new pool. By contrast, a water beetle inhabiting a more permanent habitat such as a lake will have less need to disperse and (all other things being equal) an increased investment in eggs will be repaid by greater fitness. High fecundity will therefore be traded off against a loss in flying ability.

While it would be naïve to expect that an appreciation of adaptation through optimality theory will permit us to understand all of the varied forms and behaviours of plants and animals, it does have both explanatory and predictive power (see Chapter 7: Foraging theory). Moreover, the failure of organisms to exhibit what we might expect to be the most appropriate adaptation for a particular set of ecological circumstances can sometimes shed just as much light on our understanding of the factors influencing their lives as a successful prediction.

The flexible phenotype

Though there is no doubt of the occurrence and importance of adaptations in nature, difficulties arise if we wrongly assume that all variation between individuals is inherited. Some variation is a result of phenotypic plasticity. The particular combination of genes of an individual organism (the genotype) codes, in a complex way, for the development of a phenotype, which is the total set of structural and functional properties of that organism. The phenotype is influenced by interaction with the environment during growth and development, but the expression of this plasticity is lost with the death of the organism: i.e., acquired changes in the phenotype are not passed on to the next generation — an idea that was associated with the mistaken 'Lamarckian' theory of inheritance. For example, variation in the shell shape of certain freshwater snails can be affected by a variety of factors during development, including water movement and the availability of calcium, but the precise shell shape that an adult finally acquires is not passed on to its offspring.

Our understanding of the biology of individuals and populations, and hence of community structure and ecology as a whole, is dependent upon an appreciation of the interactions between individuals and their biotic and abiotic environment. The ecology of individuals is determined by their adaptations, which are the product of natural selection operating as an optimizing process. Thus it is Darwinian natural selection which provides the organizing principle underlying ecology, and biology in general.

2

Environment and History

Hong Kong is a self-administered British Dependent Territory on the South China coast. It lies between latitudes 22°09' and 22°37'N, to the east of the Pearl River (Zhujiang) estuary (see map on front endpaper). Hong Kong consists of a section of the Chinese mainland (Kowloon and the New Territories, 782 km²) and numerous islands, the largest of which are Lantau Island (142 km²), Hong Kong Island (78 km²) and Lamma Island (13.5 km²). The total land area is 1076 km², of which 40 km² is the result of recent land reclamation. The topography is extremely rugged and there is little natural flat land. A series of ridges make up the backbone of the Territory, mostly running from northeast to southwest. The highest point is at Tai Mo Shan (957 m) in the central New Territories, followed by Lantau Peak (934 m) and Sunset Peak (869 m) on Lantau Island.

Hong Kong Island was ceded to Britain by China in 1842 and the adjacent Kowloon Peninsula was added in 1860. The New Territories, including most of the islands, were leased from China for 99 years in 1898. The entire Territory reverts to the People's Republic of China in 1997. In the 150 years since Hong Kong was founded, the population has grown from around 3000 people on Hong Kong Island, engaged mostly in farming or fishing, to nearly six million in the Territory as a whole. Because of the rugged terrain, 95% of the population live and work in less than 20% of land, and the remaining 80% is relatively undeveloped (Table 1).

In this chapter we consider those components of Hong Kong's

Table 1   Areas of Hong Kong under different land use and vegetation cover.
Data from WWF (HK).

Land use/vegetation cover	Area (hectares)	Percentage of total
Urban	17 734	16.5
Village areas	4 992	4.6
Bare rock or soil	2 115	2.0
Inland water	5 061	4.7
Mangrove	276	0.3
Other wetlands	325	0.3
Cultivation	1 414	1.3
Abandoned cultivation	3 167	2.9
Secondary woodland	10 062	9.4
Plantation woodland	4 830	4.5
Tall shrubland	11 129	10.3
Grassland with tall shrubs	8 921	8.3
Low shrubland	6 666	6.2
Grassland with low shrubs	12 902	12.0
Grassland	17 970	16.7
Total	107 564	100

physical environment which influence the ability of organisms to live where they live and to do what they do. The main factors of relevance here are Hong Kong's seasonal climate, the geology and the soils. We also describe the history of human habitation in Hong Kong, because an understanding of land-use history is essential to explain the present state of the terrestrial environment and the plants and animals which live there.

Climate and seasonality

Hong Kong has a monsoon climate, dominated by the seasonal alternation of wind direction and the resulting major contrast in weather between winter and summer (see map on end page). In winter, the continental high pressure region over Siberia and Mongolia results in north or north-easterly winds which bring cool, dry air to Hong Kong. In summer, low pressure to the north brings in warmer, moister air from over the tropical oceans. On average, the summer monsoon dominates from early May to the end of September and is replaced by the winter monsoon from November to February. Between the summer and winter monsoons are shorter periods of transitional weather. A characteristic and ecologically significant feature of Hong Kong winters is the occurrence of short-lived cold surges: outbreaks of cold Siberian air that cause a sharp drop in temperature.

The mean annual temperature is 22.8°C (1961–90) and the mean daily temperature range is 5.2°C. January is usually the coldest month, with a mean temperature of 15.8°C; July is the hottest, with a mean of 28.8°C. July is also usually the sunniest month, with a mean of 231 hours of bright sunshine, as compared with only 96 hours in March.

The mean annual rainfall is 2214 mm. Summer is the wet season, with 77% of the total annual rainfall falling between May and September, as opposed to only 6% in the four winter months. Approximately 18% of the annual rainfall — more than twice the monthly average — falls in August. Summer is also the main typhoon season and, although direct hits are rare, typhoons passing within several hundred kilometres of Hong Kong contribute about 25% of the annual rainfall in the latter part of the summer. Relative humidity exhibits a similar seasonal pattern to rainfall, except that it rises earlier in the year, during the transition period between the winter and summer monsoons. Fog is most common between February and April.

The use of the terms 'summer' and 'winter' for Hong Kong's seasons is perhaps unfortunate but well established: 'dry season' and 'wet season' might be more appropriate from the point of view of the biologist. Nevertheless, although Hong Kong is geographically within the tropics, temperature seasonality is greater than most places at a similar latitude. Indeed, the relatively large annual temperature range, cool winter, and low absolute minimum temperature, have prompted many climatologists to classify Hong Kong's climate as subtropical.

In addition to the seasonal variation in climatic means, there are considerable differences between years, particularly in the amount and timing of rainfall (Fig. 1). To a large extent these differences are a result of variation in the timing and strength of the monsoons. The long-term mean annual rainfall recorded at the Royal Observatory is 2214 mm but only 901 mm fell in 1963. The highest annual rainfall recorded was 3248 mm in 1982.

Even in a 'normal' year, monthly means of temperature and rainfall conceal brief, extreme events of ecological significance. For example, January mean temperatures hide the extreme minima brought by cold surges. The coldest period on record was in January, 1893, when during ten days of exceptional cold, ice coated the rigging of ships in the harbour. The lowest temperature reported near sea-level was 0°C

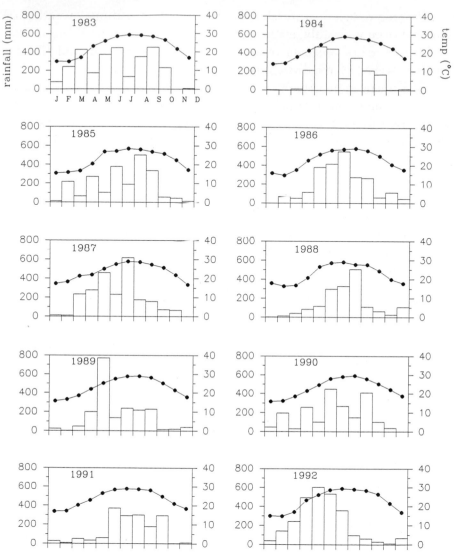

Fig. 1 Graphs of mean monthly temperature (°C) and rainfall (in millimetres) for the years 1983 to 1992. Note that the variation between years in rainfall seasonality is much greater than that for temperature. Data derived from Royal Observatory (Hong Kong government) reports.

but sub-zero temperatures probably occurred in what is now the northern New Territories. Less extreme cold events are experienced more frequently, with sea-level temperatures falling below 10°C every year and occasional frosts on higher ground and at low altitude in the northern New Territories. High-temperature extremes are probably of lesser importance in Hong Kong since the highest temperature ever recorded was 36.1°C (in August, 1990), less than 5°C above the mean daily maximum for July and August.

Extreme rainfall events are of major significance for stream ecology and can influence terrestrial communities by promoting erosion and landslides. The great majority of landslides occur during high-intensity rainfalls. The highest hourly rainfall ever recorded was 110 mm in May, 1992. Rainfall extremes are sometimes associated with typhoons, which also, of course, bring very strong winds. When a typhoon passes over or close to Hong Kong, wind speeds can exceed 120 km/h, and a maximum of 230 km/h was recorded during Typhoon Ruby in September, 1964. Winter monsoon winds and squalls associated with thunderstorms may also cause significant damage, while in June, 1982, a tornado swept across the northwest New Territories, killing two people and uprooting or breaking trees. Another hazard associated with typhoons is the storm surge. In 1937, a 6-m storm surge in Tolo Harbour destroyed several villages and killed 11 000 people.

Most of the climatic means mentioned above come from the Royal Observatory in Kowloon, which has the longest and most detailed set of records. Perhaps surprisingly, for such a small area, Hong Kong's climate also shows major variations in space. Temperature declines in a regular, predictable way with increasing altitude (approximately 0.5°C per 100 m), but the variation in mean annual rainfall is more complex. The driest place is Waglan Island, in the extreme southeast of Hong Kong waters, with a long term mean of only 1300 mm; the wettest is Tai Mo Shan, in the central New Territories, which receives over 3000 mm (see map on end page). In general, rainfall is highest on the tallest peaks and lowest in the western New Territories and the southern islands.

On a smaller scale, climate varies with slope angle and orientation (or aspect). Because Hong Kong is north of the equator, a north-facing slope receives less solar radiation than flat land or a south-facing slope, except for a period around the summer solstice (21 June). This north-south difference increases rapidly with increasing slope angle and is at a maximum in the November–January dry season. The major impact on vegetation is probably through fire, with fire frequency and intensity both likely to be higher on the warmer, drier south-facing slopes. This is in agreement with the generally greater extent of woodland and tall shrubland on north-facing slopes in Hong Kong. East-west differences in climate are less obvious but the prevailing easterly winds throughout most of the year will increase evaporation and transpiration rates on east-facing slopes, making them effectively drier. Moreover, even on a still, cloudless day, east-facing slopes will warm up and dry out earlier in the morning, increasing the chance of fire.

Hong Kong streams also undergo marked seasonal changes. Fluctuations in temperature are less than those on land, although they are greater in unshaded than shaded streams. Because stream discharge volume is determined by seasonal rainfall patterns, flows during the winter months are much lower than flood flows resulting from monsoonal rains and typhoons during the wet season. Figure 2 shows a typical example of seasonal discharge fluctuations in a small hillstream. Seasonal trends of this type are magnified by the extraction of water (for irrigation or domestic consumption) from most Hong Kong streams during the dry season. In extreme cases, the volume removed is such that surface flows of water disappear downstream of the point of extraction.

Natural seasonal variations in discharge volume are apparent from the ratio of runoff during the wet season (April to September) to that in the dry season (October to March). Data are available for seven Hong Kong catchments, and values of this 'seasonality index' range from approximately 2 to 30, indicating both a strong seasonal effect

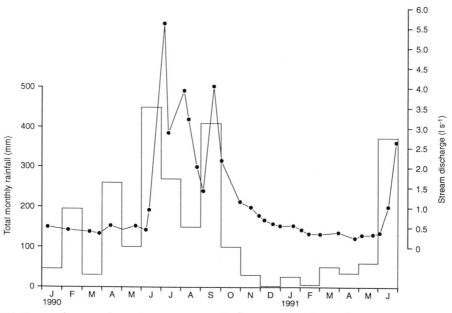

Fig. 2 Discharge volume (litres per second) of Kwun Yum Shan Stream (Sek Kong) and total monthly rainfall; January 1990 – July 1991.

as well as evidence of spatial variation in rainfall and runoff. If the maximum daily runoff or stream discharge is expressed as a percentage of total annual runoff, an indication of the intensity of spates (sudden flood flows) can be obtained. In the Shek Pi Tau basin, northern New Territories, which at 27.9 km^2 is the largest upland catchment in Hong Kong, the maximum daily flow accounts for just under 10% of annual discharge. Values for other catchments are less extreme (mean values around 4%), but nevertheless indicate the highly 'pulsed' nature of flow in Hong Kong streams.

Geology

From an ecological viewpoint, the geology of Hong Kong is fairly straightforward (see map on back endpaper). Extreme rock types, such as limestone and ultrabasics, are absent. Most of the Territory is underlain by igneous rocks of Jurassic age. These are the product of a major period of volcanic activity (the Yanshanian) during which huge volumes of lava and ash were erupted, and large granite batholiths intruded underground. The commonest rock types belong to the Repulse Bay Volcanic Group which underlie about half of the Territory. These volcanic rocks are variable but we have no evidence that any of this variation has ecological significance. The rocks consist mostly of tuffs (formed by the consolidation of volcanic ash and larger fragments ejected from the volcanoes) but include some lavas and layers of sedimentary rocks. The Repulse Bay Volcanics are mostly rhyolitic: the same as granites in chemical composition but differing in their much smaller grain size. In fact, they may have originated from the same magma as the granites, with the physical differences a result of the more rapid cooling of the volcanics. The granites, which are the second commonest rock type and cover about a third of Hong Kong, are also variable. The most common form is pink or grey and medium- to coarse-grained. Other varieties have scattered large crystals in a medium- to fine-grained groundmass.

Non-igneous rock types underlie smaller areas in Hong Kong and only the most extensively exposed are mentioned here. The Bluff Head Formation, of Devonian age, extends along the north coast of Tolo Channel and consists mainly of sandstones, conglomerates, siltstones and shales. The Lok Ma Chau Formation, of metamorphosed sedimentary rocks, underlies part of the northwestern New Territories, from Tuen Mun to Lo Wu. It is probably of Lower Carboniferous age. Finally, the Port Island Formation, of probable Early Cretaceous age, is exposed in the northeast New Territories. It consists of conglomerates, sandstones and mudstones.

Over much of Hong Kong, the bedrock is covered by superficial Quaternary deposits of various forms. Marine and beach deposits occur in low-lying coastal areas and alluvium fills the lower parts of the larger river basins. Alluvial deposition has undoubtedly been accelerated in the past millennium by deforestation of the uplands and the resulting increase in erosion. Colluvial deposits (debris resulting from landslides and other mass movements) cover large areas inland, including hill footslopes. These deposits are often several metres thick and sometimes much more. A typical colluvium consists of a soil matrix in which cobbles and boulders are embedded. No method of dating colluvium has been devised, but it is clear that many ages are present. Some of the more recent colluvial deposits may be the result of landslides triggered by deforestation.

Hong Kong is in an area of minor but significant seismic activity. Although no major earthquakes have occurred near the Territory in recent times, minor tremors are frequent. In 1874, an earthquake estimated to have had a magnitude of 5.75 on the Richter scale and an epicentre only 20 km southeast of Hong Kong caused minor structural damage.

Soils

Most hill soils in Hong Kong are formed from weathering of granite or volcanic rocks. The rate of weathering is controlled largely by the physical properties of the rock, which determine the ease of penetration by ground water. In Hong Kong's hot and wet climate, weathering is so intense and the breakdown of minerals (except quartz) so complete, that quite dissimilar rocks can produce very similar soils. The texture of the soil is, to a large extent, determined by the texture of the parent material. Since typical volcanic and granitic parent materials in Hong Kong have similar chemical composition, the often striking differences in vegetation cover and floristics between soils derived from these rock types must ultimately be the result of differences in grain size. The non-quartz components of the rock are more or less completely weathered, so it is the quartz grain size that is most significant. Thus a granite with large quartz grains will give rise to an open-textured silty-sand soil, while the typically fine-grained volcanics weather to silty-clay soils with relatively little sand. The soil texture in turn controls other important soil properties, including drainage, water retention, aeration and resistance to erosion.

The soils formed on granite and on volcanic parent materials above 500 m have been classified in the widely-used United States Department of Agriculture Soil Taxonomy as ultisols, with oxisols on volcanic materials below 500 m. This classification is, undoubtedly, a great simplification of the true picture. Moreover, it is not clear what, if any, ecological significance can be attached to this distinction, which is intended principally for agricultural use. Plant communities in general seem to respond to soil factors other than those used in the classification. With our current knowledge, the most useful division seems to be first into granite-derived and volcanic-derived soils, and then on the basis of soil depth and, perhaps, organic content. Many granitic areas are badly eroded and support only sparse vegetation (see below), while most volcanic areas have a continuous vegetation cover. Too little is known about the soils on other parent materials to draw any general conclusions, but it is perhaps significant that the insectivorous pitcher plants (*Nepenthes mirabilis*), which are otherwise confined to granite, also grow on the non-igneous Port Island Formation above Luk Keng.

Although there have been very few chemical analyses of hill soils in Hong Kong, the available information suggests that whatever the parent material these soils can be described as acid, low in organic matter, and nutrient-poor, with low to very low levels of nitrogen, phosphorous and calcium.

One conspicuous feature of soils on many Hong Kong hillsides is the extent of weathering and erosion that has taken place, producing the barren gullied badlands with rather sparse vegetation which are typical of hills in the southwest New Territories. Badlands are generally restricted to areas underlain by granite, where gulleys (as in the Tai Lam area) may be as much as 15 m deep. Granite hills in these areas tend to be rather low and rounded (compared to the higher and sharper volcanic peaks elsewhere in the Territory) with a surface cover of huge boulders exhumed as the weathered rock around them is eroded. Coarse-grained granites are more susceptible to weathering and hence erosion than fine-grained granites.

Much silt and sand is washed off badlands during torrential monsoon rains, and the eroded sediments may have damaging effects on streams where they clog, smother or bury the natural substratum. Such clogging may interfere with natural drainage patterns and lead to waterlogging or flooding in lowland areas. Under such conditions, impounded drainage waters may sometimes carry enough iron leached from granite soils to make the water toxic to plants and animals. The chemistry of iron in freshwater is extremely complicated, but where dissolved oxygen is freely available, iron is precipitated from solution as an apparently non-toxic complex. These complexes form the reddish flocculant deposits seen in small streams draining waterlogged lowlands and abandoned paddy fields.

Environmental history

Hong Kong lies on part of the South China coast which has been relatively stable tectonically, so changes in the geography of the Hong Kong region over the last million years have been largely the result of global changes in sea-level. Sea-levels fell during the glacial periods (Ice Ages) and rose to near present levels during the warmer interglacials. The maximum fall during the last glacial period occurred about 15 000 years ago, when sea-levels were 100 m below present levels leaving Hong Kong more than 150 km from the coast. Sea-levels rose to their current height some 6000 years ago. Subsequent alterations to the coastline have been the result of sedimentation. Most of the low-lying land in the Pearl River delta seems to have been formed in the last few thousand years by this process, although reclamation has been locally important.

While the basic topography of upland Hong Kong was unaffected by sea-level changes and sedimentation, the flora and fauna of the hills and streams has changed dramatically. Two factors are responsible for this: climatic change and human impact. There is still very little direct evidence concerning past climates in the Hong Kong region. Limited fossil evidence (mostly from pollen preserved in estuarine and marine sediments) and climatic modelling suggest that Pleistocene climates were somewhat cooler and considerably drier than the present day. At the glacial maxima, the broad-leaved evergreen forests of the interglacials may have been replaced by a more open, pine-dominated woodland, but this is largely speculation. We can be certain, however, that broad-leaved forests were established once again at the end of the last glacial episode, 10–12 000 years ago. Although the climate has continued to fluctuate since then, the major environmental changes that have occurred have been entirely the result of human impact.

It is conventional to think of major human impact on the natural world as being an event of the last few thousand years only. Yet humankind, in the form of *Homo erectus*, has been in Asia for at least one million years. We know little about the ecology of early humans, but they must have been skilled hunters and were able to spread widely across Africa and Asia. *Homo erectus* probably had spears and may have controlled fire, although this is still a controversial matter.

Whatever the influence of *Homo erectus*, there is no doubt that the most drastic impact on Hong Kong's landscape can be attributed to modern humans. There is fossil evidence for the presence of early *Homo sapiens* in Guangdong from at least 140 000 years ago but the first archaeological records for Hong Kong are much more recent. The earliest known settlements (dating from the fourth millennium B.C.) were coastal, as at Yong Long (near Tuen Mun) where recent excavations have provided evidence of one site which was populated between 4000 and 2500 B.C. Since all of the older coastline is now submerged, the absence of even earlier evidence of human settlement is not surprising. Scattered excavations give us no information on human population densities so we have no idea of what impacts, if any, people were then having on inland habitats.

We do know that sea salt was exported from Xin An County (which included all of modern Hong Kong) during the Han Dynasty (206 B.C. – A.D. 265), and pearls were collected from the sea near Tai Po during the Tang Dynasty (eighth century). From archaeological evidence, it is clear that kilns for the production of lime from shell and corals were established at virtually every suitable site on the coast in the period A.D. 300–900. Again, we do not know what the implications are for human population densities in Hong Kong during this period. The lime, like the salt and pearls, was probably for export but the fuel requirements must have had at least a local impact on the surrounding forest. However, we have no idea how many of these kilns were operated at once. No settlement from this period has yet been excavated.

There is some evidence for settlement in the present Kowloon City area as well as elsewhere in the New Territories from the Song Dynasty (960–1278), and there are well-established records of settlement from the fourteenth century during the Ming Dynasty (1368–1644), when incense wood was exported. Genealogies of the major New Territories lineages support this history although there is a lack of corroborating evidence to substantiate their claims of settlement. Some of the genealogies may have been exaggerated to enhance land claims and, in any case, they do not provide evidence for more than a few hundred people. This does not mean that there was not extensive agricultural settlement in Hong Kong by this period or even considerably earlier, just that we have little evidence for or against this.

Genealogies and relics from the early Qing Dynasty (1644–1912) are relatively abundant, and it is clear that in the last few decades of the Ming the Punti (i.e., Cantonese) lineages from northern Guangdong had established quite large populations in the Hong Kong region. Lineage members were linked by patrilineal descent, and ownership of land was the focus of lineage organization. Much of their colonization of the region probably took place during the fifteenth and early sixteenth centuries, but the Tang lineage settled at Kam Tin during the late Song Dynasty in the thirteenth century. The first ancestral hall in the New Territories was built in 1525 by the Tangs, while Yuen Long Market was founded in the middle of the seventeenth century, followed by Tai Po Market in 1672.

The seventeenth century was a period of banditry and disorder, and famine resulted from these disturbances. The unrest culminated in a period known as the 'coastal evacuation' (1662–69) when, as part of measures against Ming loyalists and pirates, the emperor ordered the inhabitants of coastal Guangdong (including the New Territories) to move 50 *li* (approximately 28 km) inland. Military guard posts were established in the depopulated New Territories. Although it is unlikely that the evacuation was complete, the process caused much hardship.

Significantly, none of the existing Hakka lineages in Hong Kong trace their residence in the New Territories back to before the coastal evacuation. This is important because, among the ethnic groups making up the New Territories inhabitants, the Tanka and Hoklo were boat dwellers, the Cantonese inhabited valley bottoms and lowlands, while the Hakka cleared and cultivated hillsides and other marginal land leased from the Cantonese. Hakka settlement must have had a major impact on the upland landscape. In essence, certain well-established Cantonese lineages claimed the so-called 'bottom-soil rights' to much of the Territory, and rented top soil to others. Topsoil 'owners' were perpetual lessees paying a rental to the subsoil, or true, owners. The 'one field — two lords' system may have been established in Hong Kong as early as the late Song Dynasty in the thirteenth century, and the principal land owners, the Tang family of Kam Tin, held at least part of Hong Kong Island on this basis before 1573.

The general pattern appears to have been that the Cantonese settled in the most productive areas from before the Qing Dynasty, while the Hakkas moved in from the late seventeenth century and cultivated the uplands and other marginal areas. The ecological impact of the Hakka immigration cannot be assessed accurately in the absence of information on the seventeenth-century vegetation and biota, but the establishment of numerous Hakka villages in areas which were apparently only sparsely inhabited before must have been of major significance. It therefore seems most likely that the seventeenth century saw the final transformation of what was once a forested landscape with some cleared areas into the barren hills reported by European visitors during the nineteenth century.

Agriculture recovered quickly from the disturbance of the coastal evacuation, and export of incense wood was resumed. However, periods of privation still occurred due to the vagaries of a destructive climate, and disturbances resulting from feuds between local inhabitants and attacks by bandits and pirates. Resumption of rice growing and extensive cultivation of the hills after the coastal evacuation was accompanied by the widespread practise of stream-course modification and flow regulation. In particular, water was diverted into irrigation channels established to serve agricultural lands. Stream banks were often strengthened with stone diking, to prevent erosion of surrounding fields and to facilitate maintenance of irrigation systems.

Even the highest hills were not spared human impact: the Xin An (Sun On) Gazetteer of 1688 refers to tea cultivation on Tai Mo Shan, Castle Peak and Lantau Peak. Old tea terraces can also be seen on other high slopes in the New Territories and on Lantau, confirming that human impact during this period was not confined to the fertile lowlands. Indeed, Tai Mo Shan was famous for the production of green 'cloud and mist tea', and was terraced to near the summit. The age of the tea terraces is not known but they probably predate the Hakka settlements of the seventeenth and eigtheenth century. Upland areas which were not terraced were important as a source of grass and shrubs for fuel and were grazed by draught cattle which roved the hillsides when not at work. Collectors of herbs, trappers of birds, snakes and other wild animals, and those who combed hillstreams and pools for items that served medicinal purposes, must also have had a significant impact on the upland flora and fauna.

The earliest description by a visitor from Europe of part of the present Territory of Hong Kong comes from the great Swedish botanist, Peter Osbeck, who sailed past Lantau Island on his way up the Pearl River estuary to Canton (Guangzhou) in 1751. His account of the voyage has the following passage, 'We had Lantoa on our right and the southern isles of Limes [the Lema Islands] on the left: the sea formed high billows rolling in from the isles, which were quite green with plants, but had no woods.' If, as this passage suggests, the steep and rugged hills of Lantau Island had been deforested by this time, it is likely that the rest of the present Territory had suffered the same fate.

From the 1840s onwards, many visitors described the landscape of the new colony on Hong Kong Island. Adjectives such as 'barren', 'bare', 'bleak' and 'sterile' occur in all these accounts. The hills were covered in coarse grass, with a few scattered shrubs and stunted pine trees. Cultivated land was confined to the lower parts of the larger valleys, where sweet potato was the major crop. Visitors with botanical training soon discovered, however, that first impressions were deceptive. As George Bentham remarked, in the introduction to his *Flora Hongkongensis* (1861), 'one is struck with the very large total amount of species crowded upon so small an island, which all navigators depict as apparently so bleak and bare'.

Where were all these species? Early descriptions and the earliest maps mention only three areas with substantial woodland: Happy Valley, Little Hong Kong (Aberdeen), and Tai Tam Tuk. We know virtually nothing about the woodland at Tai Tam Tuk: neither its flora nor its eventual fate. The Happy Valley Woods on the slopes behind the present race course were visited by many botanists, so we know a lot about the flora, which was extremely rich and included some species recorded nowhere else in Hong Kong. The Little Hong Kong Woods were the most extensive and the only one of which a remnant still survives, north of Nam Fung Road, around the Hong Kong Electric tunnel portal. Outside these woodlands, much of the flora was confined to damp ravines and other similar sites, protected from fire and less accessible to woodcutters.

In contrast to the botanists, nineteenth-century visitors with zoological interests were unimpressed. The great ornithologist, Robert Swinhoe, a visitor to Hong Kong in 1860, remarked that he saw only one species of bird 'which does not occur somewhere in the neighbourhood of Amoy [Xiamen]', and went on to state that, relative to Hong Kong, Canton (Guangzhou) 'literally swarms with birds'. Marine biologist William Stimpson, who visited Southern China in 1854, had little to say about terrestrial environments, but noted the presence of deer (presumably barking deer, *Muntiacus reevesi*) on Hong Kong Island and rhesus macaques (*Macaca mulatta*, which he termed 'baboons') on the Lema Islands. Stimpson also collected land and freshwater snails, some of which were later revealed as species new to science.

Nineteenth-century descriptions of what later became the New Territories echo the early accounts of Hong Kong Island but give few details. The first reasonably complete description is that in Stewart Lockhart's report to the Colonial Office on the extension of the Colony in 1898. In comparison with Hong Kong in 1841, the New Territories in 1898 had a much larger population (around 70 000) and much more cultivated land. Moreover, the New Territories' economy had already been strongly influenced by the proximity of the new urban centre and port around Victoria Harbour. Rice was by far the most important crop. Other crops included sugar cane, indigo, hemp, peanuts, various root crops, sesame, vegetables and fruit. Lockhart saw rice growing up to 420 m, and both tea and pineapples at 460 m on the northern slopes of Tai Mo Shan. There were no extensive forests but pine trees (*Pinus massoniana*) were grown for firewood on some of the lower hill slopes. Lockhart also mentioned the presence of 'thick clumps of well-grown trees' around many of the villages. Other accounts from this period suggest that at least some of these 'clumps' were, in fact, small woodlands, floristically similar to the one at Little Hong Kong. As on Hong Kong Island, trees also survived in sheltered ravines, particularly at high altitude. Most of the hills, however, were covered in grass or ferns, and large areas were cut annually for use as domestic fuel, as still happens today in rural Guangdong.

The initial ecological impact of the new colony must have been entirely negative, both on Hong Kong Island itself and on those parts of the mainland which supplied food and fuel to the growing settlement. Although early attempts were made to control the cutting of trees, shrubs and grass for fuel on Hong Kong Island, these seem to have had little, if any, effect. The first positive impacts resulted from the beginning of afforestation. Extensive planting started in the early 1870s and peaked in the 1880s, when a million or more trees were planted out or sown *in situ* every year. Most were native pines (*Pinus massoniana*) but smaller numbers of other species, both native and exotic, were included.

By 1900, most suitable land had been planted on Hong Kong Island, and planting started in the New Territories. The native pine was still preferred for the most difficult sites, but the plantings during this period included many rare native trees and shrubs, with the result that the original distributions of these species within the Territory have often been obscured. The diversity of the forest at Tai Po Kau, for instance, owes more to man than to nature. By 1938, 70% of Hong Kong Island was covered in government plantations and there were several large afforested areas (including Tai Po Kau Forest) in the New Territories. An additional 200 km² of the New Territories was occupied by leased forest lots, mostly covered in a sparse growth of pine trees, used as a source of firewood. The government plantations, in contrast, were now viewed largely as protection forests rather than a source of timber or fuel.

Fifty years of forestry work was destroyed between 1940 and 1946. Extensive felling to provide firewood for Hong Kong started in 1940 as fuel supplies from China were cut off by the war. Cutting accelerated during the Japanese Occupation, from December 1941 to August 1945, and continued in 1946, after liberation and until sufficient fuel could be imported. The first post-war reports speak of almost complete destruction of the forests in the Territory. Aerial photographs from 1945, however, show that many village *feng shui* woods (Chapter 5) survived with their canopies, at least, intact. There were also small wooded areas in protected valleys at high altitude. Hong Kong was once again a 'barren rock' with hidden pockets of diversity.

During the Japanese Occupation, Hong Kong's population was reduced from about 1.6 million to 0.6 million (by a combination of mass deportations and voluntary returns to China), but a huge influx from across the border brought the total to 2.2 million by 1950. Since then, it has increased by (very roughly) 1 million per decade and, today, it is approximately 6 million. The population influx during the 1950s, coupled with a loss of the entrepot trade when China entered the Korean War, spurred a rapid transition to a manufacturing economy. Development proceeded at a tremendous pace, with new industrial and housing estates, new reservoirs to meet the increased demands on the water supply, and large-scale reconstruction of the road network. During the last three decades, there has been a marked movement of both population and industry away from the previous concentration around Victoria Harbour into outlying areas, at first on Hong Kong Island and in New Kowloon and then, increasingly, into New Towns (and expanded old towns) in the New Territories. Much of the new development has taken place on land reclaimed from the sea.

The post-war population growth and rapid industrialisation have affected all aspects of the Hong Kong landscape. There have been major changes in the pattern of farming, with a decline in the area used for rice (culminating in the disappearance of rice paddy from the Territory) and an increase in the intensive cultivation of vegetables and other high-value crops. Over the same period, much former paddy land around remote villages was abandoned and the villagers themselves moved into urban areas or emigrated.

More recently, there has been a progressive encroachment of non-agricultural impacts upon lowland areas, particularly in the central and western New Territories. Competition with cheaper produce imported from China, where labour costs are lower, has resulted in the abandonment of agriculture even in relatively accessible places. Unfortunately, the abandoned fields have not reverted to wild lands but instead have become container parks and scrapyards.

The cutting of hillside vegetation for domestic fuel declined rapidly after the war as alternative fuels became cheaper and widely available. Secondary shrubland and forest developed on some hillslopes, particularly on Hong Kong Island, although periodic fires maintained others as grassland. Afforestation efforts increased but never attained the scale they had before the war. Tree planting was concentrated in the new water-catchment areas and was justified on grounds of reduced erosion. Illegal hunting and trapping of wildlife in the late 1970s led to fears that no large vertebrates would escape extinction, but these practices seem to have declined during the last decade.

Hong Kong is now in the middle of another economic transition, as manufacturing industry moves across the border and the Territory becomes a service centre for an increasingly prosperous hinterland in South China. Rising living standards have brought both costs and benefits for the Hong Kong environment. The costs include a partial reversal of rural depopulation as new roads, improved public transport, and wider car ownership make it possible to live in areas considered too remote a decade ago. Increased mobility, combined with more leisure time, has also resulted in greater recreational impact on the countryside on weekends and public holidays. The benefits include a serious attempt to impose planning controls on new developments, and an increasing public concern for Hong Kong's obvious and burgeoning environmental problems.

3

Climate and the Hong Kong Biota

Is Hong Kong tropical? The simplest answer is yes, it is more than 100 km south of the Tropic of Cancer and thus well within the tropics. However, if the question is 'Does Hong Kong have a tropical climate?', the answer is less obvious. Hong Kong's climate has features which are not typical of the tropics as a whole. As noted in Chapter 2, temperatures below 10°C — in the range known to cause chilling damage to sensitive plant species — occur at least a few days every winter, while temperatures below 5°C are recorded several times each decade. A sea-level frost (0°C) has been recorded only once on Hong Kong Island (in 1893) but frosts occur more frequently at higher altitudes and in the northern New Territories. Both climatologists and plant geographers have tended to define the tropics in a way that excludes Hong Kong, placing it instead in the subtropics or an intermediate 'transitional tropics' zone. In marked contrast, zoogeographers have had no doubts that Hong Kong is well within the tropics. For example, the vast majority (99%) of the 202 butterfly species recorded from Hong Kong originate in the Oriental tropics; only three are species with a primarily north-temperate range. The moth fauna, too, is dominated by tropical representatives. Similarly, of the 21 bat species found locally, only one (the noctule, *Nyctalus noctula*) can be described as being temperate or non-tropical.

The different views of plant and animal geographers probably reflect real differences in how plants and animals respond to brief periods of extreme (for the tropics) cold. Plants are immobile and mostly long-lived: in the above-ground parts, at least, minimum cell temperatures are likely to equal minimum air temperatures. Only buried seeds and underground storage organs can escape. In contrast, most terrestrial animals are mobile and can move to favourable microhabitats, while birds and mammals can thermoregulate by using more energy to maintain body temperature. The temperature of streams is buffered by heat from the soil, with only the very smallest bodies of water freezing even at high altitudes.

In most parts of the world, there is a belt of drier climates, with non-forest vegetation, between the forests of the tropics and those of the temperate zone. Only in East Asia was forest cover, at least until the human disturbance of the past millennium, continuous from near the equator to the Arctic treeline. The absence of a sharp break in the vegetation (or in the associated fauna) makes it particularly difficult to say where the tropics ends and the subtropics begins. In any case, arguments about where to place a boundary along a gradient of continuous, gradual change are neither interesting nor informative. The real question is how Hong Kong's biota differs from that of areas to the north and south and to what extent these differences can be explained by differences in climate.

Vascular plants

The majority of Hong Kong's flora consists of genera in which most species are found within 10–15° of the equator, and only a minority of genera have their predominant distribution to the north. That is to say, Hong Kong's flora is largely tropical at the generic level. However, several major, strictly tropical families of Asian plants do not penetrate as far as Hong Kong (e.g., Burseraceae, Dipterocarpaceae, Myristicaceae), and a number of largely extra-tropical families or genera (e.g., Ericaceae, *Machilus* and *Ilex*) are better represented here than in the lowland tropics further south. Hong Kong's flora is tropical, but less so than places nearer the equator. The best comparison is with the mid-altitude forests of equatorial mountains, where the families Fagaceae, Lauraceae, Myrtaceae, and Theaceae are also prominent, while lowland families, such as the Burseraceae, Dipterocarpaceae and Myristicaceae, are rare or absent. But it must be borne in mind that, while mean annual temperatures on equatorial mountains are similar to those in Hong Kong, the montane climate resembles the equatorial lowlands in its lack of seasonality.

The probable influence of cold extremes on plant distributions in Hong Kong was graphically illustrated by the impact of the exceptional low temperatures which occurred on 28 December 1991, when a surge of cold air from the north produced temperatures below 5°C all over the Territory and sub-zero temperatures at high altitude. On Tai Mo Shan the absolute minimum recorded (at about 850 m) was –4.7°C, making it the coldest day in 15 years. Light rain resulted in ice formation on vegetation from 480 m upwards. From all the available evidence, it seems likely that sub-zero minimum temperatures occurred down to about 400–450 m on Tai Mo Shan and that, over the whole Territory, the 'frost line' varied between 250 and 550 m.

The impact of the cold on plant species was highly selective and varied with altitude. At sea-level, where the minimum temperatures recorded during the cold event ranged from 2.1°C (Tuen Mun) to 4.9°C (Kai Tak Airport), very few native plant species were visibly injured. The most conspicuous damage was to the small coastal tree, *Macaranga tanarius*, which is also widely planted inland. Typically the youngest and oldest leaves suffered most but the extent of injury varied greatly and some individuals appeared unscathed. At Mai Po Marshes, some newly-expanding leaves of the mangrove tree, *Avicennia marina*, were killed, particularly on individuals growing at the seaward margin of the mangrove forest. Significantly, both *Avicennia* and *Macaranga* in Hong Kong are near the northern limits of their vast, tropical ranges.

Above 250 m on Tai Mo Shan, but still below the frost line, some damage to native, non-coastal species was visible. Two fig species, *Ficus hispida* and *F. fistulosa*, a small tree, *Aralia spinosa*, and a fern, *Blechnum orientale*, seemed to be the most cold-sensitive. Widespread but highly-selective damage to native species started about 450 m, just above the probable frost line. With increasing altitude, the degree of damage to sensitive species (such as those listed earlier) increased and more species were affected, although the majority still showed no visible injury. Towards the summit of Tai Mo Shan (957 m), most of the more sensitive species are absent and signs of frost injury began to appear in some species that were undamaged at lower altitudes. However, even in the highest and most exposed positions, many species were completely unaffected and some were in flower a week later (e.g., *Eurya chinensis*, *Daphne championii* and *Pittosporum glabratum*).

The pattern of frost damage among plant taxa was not random. All the more frost-sensitive woody plants are members of genera in which the majority of species occur in the tropical lowlands, south of

Hong Kong. All of the most frost-tolerant woody species, in contrast, are in genera with extensive extra-tropical distributions and which are typically rare or absent in tropical lowlands. The neatness of this pattern is somewhat obscured when species of intermediate tolerance are looked at, but it is striking none the less. For most woody species, frost damage was confined to the leaves and young shoots and was probably only a temporary set-back. However, the differential susceptibility of species from tropical and extra-tropical genera strongly suggests that low temperatures are a significant biogeographical and ecological factor in Hong Kong. This is supported by the altitudinal zonation of the flora, with tropical genera (such as *Aporusa*, *Ficus*, *Melastoma*, *Psychotria* and *Rhodomyrtus*) increasingly replaced by extra-tropical genera (such as *Camellia*, *Castanopsis*, *Eurya*, *Ilex* and *Quercus*) from about 500 m upwards.

The great frost of January 1893 produced similar, but more severe, effects on the vegetation of Hong Kong Island. The greater degree of damage may, in part, have been the result of the longer period of cold, with at least four days of frost at higher altitudes. Unfortunately, there are no records of temperatures or vegetation damage from the area which is now the New Territories, where even lower temperatures must have occurred. Less extreme cold periods, of the type which occur annually, cause no visible damage to the vegetation, even at high altitudes. Presumably, those species poorly adapted to withstand cold conditions have already been eliminated or are restricted to the lowlands.

Fauna

Low temperatures would be expected to affect birds and mammals differently from plants. They can maintain their body temperature by increasing their metabolic rate. Increasing the metabolic rate, however, requires increased food consumption. Even if excess food is available, there is apparently a maximum metabolic rate that any particular organism can sustain, so cold temperatures requiring higher metabolic rates will be fatal. In practice, limits to food availability are likely to limit a wild animal's ability to survive prolonged cold before this absolute physiological limit is reached.

Although we have no evidence that low-temperature extremes ever kill mammals in Hong Kong, there are reports of birds dying in cold weather. The bird deaths have been of winter visitors from the north — red-flanked bluetails (*Tarsiger cyanurus*) at Tai Po Kau and yellow-browed warblers (*Phylloscopus inornatus*) at Mai Po — but even typical winter temperatures in Hong Kong must be stressful for animals adapted to a warmer climate. Unlike equatorial birds and mammals, which never need to spend additional energy to maintain body temperature, low temperatures must significantly increase energy demand in the coldest months in Hong Kong. This will be particularly true for small-bodied species because of their relatively large surface-to-volume ratio. In addition, and probably more importantly, temperature seasonality in Hong Kong contributes to major fluctuations in food availability for all but the most unselective feeders. For insectivores, these two factors must reinforce each other, because the winter maximum in energy demand coincides with a minimum in the availability of insect food. Indeed, horseshoe bats (*Rhinolophus* spp.) enter a torpid, 'dormant' state during the winter, although other insectivorous bats such as the bicoloured roundleaf bat (*Hipposideros pomona*) remain active throughout the year.

Despite the winter chill, Hong Kong's bird and mammal faunas are clearly tropical. Many bird and mammal families have virtually worldwide distributions but others are largely concentrated in the tropics and several of these occur in Hong Kong. Among our resident birds, the barbet (*Megalaima virens*), bulbuls (*Pycnonotus* spp.), drongos (*Dicrurus* spp.), flower-peckers (*Dicaeum* spp.), laughing-thrushes (*Garrulax* spp.), munias (*Lonchura* spp.), shrikes (*Lanius* spp.), scarlet minivet (*Pericrocotus flammeus*) and fork-tailed sunbird (*Aethopyga christinae*) are primarily tropical. Civets (*Paguma larvata* and *Viverricula indica*), mongooses (*Herpestes* spp.), fruit bats (*Cynopterus sphinx* and *Rousettus leschenaulti*), and the Chinese pangolin (*Manis pentadactyla*) are tropical mammals although they also penetrate far into the subtropics. As with the flora, some families which are typical of the Asian tropics are absent but, in this case, it is more likely to reflect direct and indirect human impact than climatic constraints. The primeval forests that once covered Hong Kong probably supported flying squirrels, gibbons, hornbills and other taxa now confined to Hainan and the extreme south-west of China. Surprisingly, few, if any, birds and mammals in Hong Kong come from groups with largely extra-tropical distributions. Moles, weasels and martens are absent, although a few representatives of these groups penetrate into tropical southeast Asia. Moreover, the South China red fox (*Vulpes vulpes hoole*) and the South China badger (*Meles meles leptorynchus*), which used to occur in the territory (the former died out recently and the latter was recorded last in 1922), are at the southern limit of their range in Asia and seem to have undergone some morphological differentiation from more northern populations; the southern race of badger, for example, is smaller than specimens from temperate localities, and the pelt is slightly different.

The effects of short periods of cold on reptiles and amphibians are unlikely to be severe as most of them hibernate or become inactive in protected lairs during the winter. However, if chilling occurred at a time when they were active and breeding, then the effects of unusual cold on reptiles and amphibians would be devastating. Many insects, likewise, pass the winter in a diapause or resting stage. Eyewitness reports indicate that those insects which were abroad during the great frost of 1893 were killed by the low temperatures. Populations of insects did, however, build up again in the spring of that year although a few butterflies that emerged from their pupae soon after the cold spell were reported to show signs of deformation. There was no apparent long-term effect of the December 1991 chill on terrestrial animals, or those inhabiting streams. A lack of effect on stream animals reflects the fact that stream temperatures are buffered during short cold snaps. During most winters, average daily stream temperatures do not fall below 14°C. Even during the 1991 chill, when air temperatures around a small stream in the New Territories fell to 1.3°C, water temperatures did not drop below 12.5°C.

The stream fauna of Hong Kong is predominately tropical, with few temperate elements. Crustaceans such as *Gammarus* (Amphipoda) and *Asellus* (Isopoda), which are characteristic of north-temperate streams, are absent and replaced by freshwater crabs and shrimps which do not occur in streams outside the tropics and subtropics. Members of the snail family Thiaridae which are abundant in Hong Kong streams are not found in temperate latitudes. Likewise, fishes in the family Balitoridae (previously assigned to the Homalopteridae by some systematists) — which are common in the Territory — are confined to tropical Asia, and the Belontiidae (represented locally by the paradise fishes, *Macropodus opercularis* and *M. concolor*) reach the northern limit of their range in South China. The composition of major groups of aquatic insects also illustrate the tropical character of our stream fauna. Among Trichoptera (caddisflies), the Limnephilidae which are highly diverse in temperate streams, are completely absent from Hong Kong and are replaced to some extent by the Calamoceratidae — a typically warm-water family. The same pattern is seen among the Stenopsychidae: they have speciated widely in the streams and rivers of northern Asia, but only a single species has been found in tropical Hong Kong. Similarly, several species-rich families of stoneflies (Plecoptera) occur in the streams of northern China, and at least eight families are found in Japan, yet no more than three are known from Hong Kong, one of which (the Leuctridae) is represented by a single species only.

Overall, it appears that short periods of low temperatures during the cooler months of the year have little effect on the local fauna and flora. This may be because natural selection acting over many generations has adjusted seasonal patterns of activity so that the winter is spent as a dormant or resting stage. Alternatively, it may reflect the differential ability of members of the pool of available species in the region to survive in Hong Kong; i.e., only species with appropriate tolerance to cold are able to persist in Hong Kong. Regardless of which explanation is correct (and they are not mutually exclusive), it is evident that many members of the biota must make seasonal adjustments to their behaviour and metabolism in order to deal with the local climate. It is appropriate at this stage, therefore, to examine the patterns of animal and plant seasonality in Hong Kong.

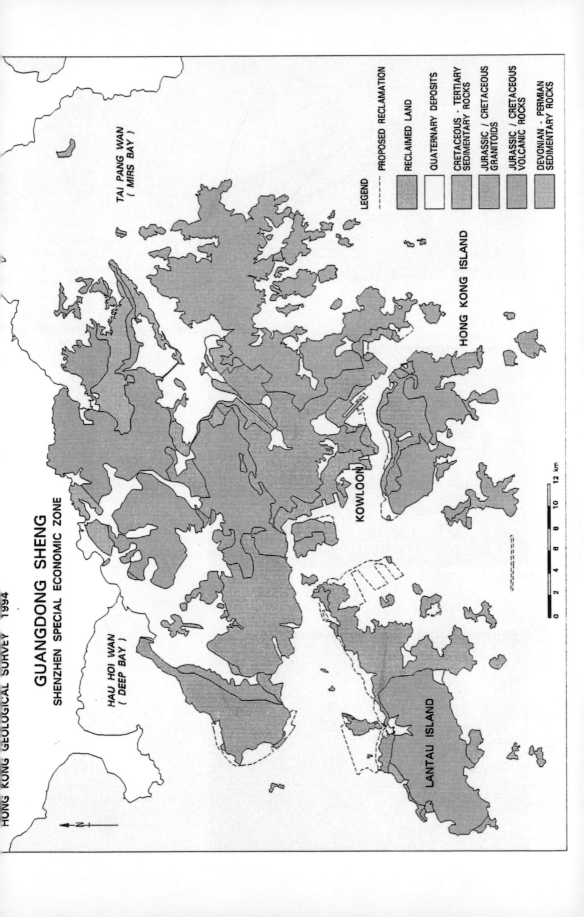

GUANGDONG SHENG
SHENZHEN SPECIAL ECONOMIC ZONE

HAU HOI WAN
(DEEP BAY)

TAI PANG WAN
(MIRS BAY)

KOWLOON

HONG KONG ISLAND

LANTAU ISLAND

LEGEND

- - - - PROPOSED RECLAMATION

RECLAIMED LAND

QUATERNARY DEPOSITS

CRETACEOUS - TERTIARY
SEDIMENTARY ROCKS

JURASSIC / CRETACEOUS
GRANITOIDS

JURASSIC / CRETACEOUS
VOLCANIC ROCKS

DEVONIAN - PERMIAN
SEDIMENTARY ROCKS

0 2 4 6 8 10 12 km

N

Plate 1
Remains of tea terraces close to the summit of Tai Mo Shan. Photo by David Dudgeon.

Plate 2
The atlas moth (*Attacus atlas*: Saturniidae), with a wing span of 210 mm, is a tropical representative of Hong Kong's moth fauna and one of the largest moths in the world. Photo by Michael Lau.

Plate 3
Nyctalemon menoetius, a large (wing span 110 mm) and striking moth of the exclusively tropical family Uraniidae feeding at a flower cluster of *Cleistocalyx operculata* (Myrtaceae). Photo by Gary Ades.

Plate 4
The small Indian civet, *Viverricula indica*, feeds on rats, other small vertebrates, large insects and fruit. It is an important seed dispersal agent. Photo by Michael Lau.

Plate 5
The fruit bat, *Rousettus leschenaulti*, feeds on a wide variety of fruits but is particularly fond of figs. Photo by Gary Ades.

Plate 6
The paradise fish, *Macropodus opercularis*, inhabits streams and marshes where the male builds a floating bubble nest for breeding. The female is driven away by the male after the eggs are laid in the nest, and he remains to guard the eggs and newly hatched young. Photo by Chong Dee-hwa.

Plate 7
The wingless litter cockroach, *Opisthoplatia orientalis* (body length 4 cm), is common among fallen leaves and wood in forests and shrubland in Hong Kong. It plays an important ecological role as a consumer of leaf litter, and is preyed upon by many animals including civets. Photo by Chong Dee-hwa.

Plate 8
A carpenter bee, *Xylocopa iridipennis*, visiting *Lantana camara*. These large bees are probably the major pollinators of several common shrubland plants, including species of *Melastoma. Lantana,* however, is an exotic. Photo by Chong Dee-hwa.

Plate 9
The Chinese bulbul, *Pycnonotus sinensis*, is the most important seed dispersal agent in shrubland and secondary forest. Photo by Chong Dee-hwa.

Plate 10
The crested bulbul, *Pycnonotus jocosus* — photographed here on Chinese hackberry, *Celtis sinensis* — has similar habits to the Chinese bulbul but is more common in urban areas and avoids woodland. Photo by Chong Dee-hwa.

Plate 11
The white-eye, *Zosterops japonicus*, is another important seed dispersal agent, but its small size means it is unable to swallow some of the larger fruits in shrubland and secondary forest. Photo by Michael Lau.

Plate 12
The rubythroat, *Luscinia calliope*, is a common winter visitor to Hong Kong. Although largely insectivorous it also eats a variety of fruit. Photo by Gary Ades.

Plate 13
The spotted munia, *Lonchura punctulata*, is usually seen in small flocks and feeds mainly upon unripe grass seeds. It is especially fond of wild millet (*Echinochloa crusgalli*) and the exotic Guinea grass (*Panicum maximum*). Photo by Chong Dee-hwa.

Plate 14
The black drongo, *Dicrurus macrocerus*, is an insectivorous summer visitor, which takes large insects by sallying from a perch overlooking open ground — especially (as here) abandoned cultivation. In the evening these birds can sometimes be seen around street lamps feeding on moths which are attracted to the light. Photo by Chong Dee-hwa.

Plate 15
The black kite, *Milvus migrans*, is Hong Kong's commonest bird of prey. Despite its fierce appearance, it is largely a scavenger. Photo by Chong Dee-hwa.

Plate 16
The peregrine falcon, *Falco peregrinus*, feeds on birds which it kills with its feet by diving down upon them at great speed. The specimen shown here was rescued from a local market, and the restraints upon the legs are a temporary measure while the animal is rehabilitated before release. Photo by Chong Dee-hwa.

Plate 17
A roosting group of insectivorous mouse-eared bats, *Miniopterus* spp. Photo by Gary Ades.

Plate 18
Pallas' squirrel, *Callosciurus erythraeus thai*, is common on Hong Kong Island. This subspecies is believed to have been introduced from Thailand. Photo by Chong Dee-hwa.

Plate 19
The spectacular lantern bug, *Pyrops candelaria*, is abundant on litchi, longan and citrus trees in late March and April. The function of the conspicuous upturned 'snout' is not known. Photo by Chong Dee-hwa.

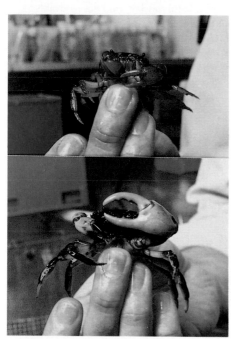

Plate 20
The freshwater crab, *Somanniathelphusa zanklon*, showing a mature female (upper) and a male (lower) with a particularly well-developed left chela (claw). Photos by David Dudgeon.

Plate 21
Nanhaipotamon hongkongense is Hong Kong's most striking freshwater crab. This upland species is perhaps better described as amphibious because it excavates burrows in damp areas on land, and runs about freely during periods of heavy rain. Photo by Chong Dee-hwa.

Plate 22
Cryptopotamon anacoluthon
feeds on the terrestrial leaf litter
that falls into Hong Kong
streams. Other freshwater
crustaceans, such as the shrimp
Neocaridiana serrata, have
similar habits. Photo by Chong
Dee-hwa.

Plate 23
Euphaea decorata is perhaps
Hong Kong's commonest and
must widespread stream
damselfly (Odonata). The
larvae are aquatic predators
while the adult males (shown
here) hold territories around
streams within which they
compete for females. Photo by
Chong Dee-hwa.

Plate 24
The bright colours of
Rhinocypha perforata males
make these insects conspicuous
whilst they occupy territories.
Territorial behaviour in this
damselfly is usually centred
around a large partially-
submerged rock in midstream;
the rock is used as a perching
site when the insect is not in
flight. Like *Euphaea decorata*
(Plate 23), *R. perforata* is
frequently encountered along
Hong Kong hillstreams. Photo
by Chong Dee-hwa.

Plate 25
Ctenogobius duospilus: the colourful dorsal fin and patterned gill covers of the males of this common goby are used in displays and contests over ownership of patches of territory on the stream bed. Photo by Chong Dee-hwa.

Plate 26
The Hong Kong newt, *Paramesotriton hongkongensis*, is protected by law locally, and is easily recognized by the distinctive body shape and red markings on the abdomen. Photo by Michael Lau.

Plate 27
The three-banded box terrapin, *Cuora trifasciata*, is found in upland streams but has become scarce locally because consumption of the flesh is thought to bring medicinal benefits. It is now protected by law in Hong Kong. Photo by Michael Lau.

Plate 28
The Chinese big-headed terrapin, *Platysternon megacephalum*, is an uncommon reptile found in and around streams on Tai Mo Shan and Ma On Shan. Photo by Gary Ades.

Plate 29
The masked palm civet, *Paguma larvata*, plays an important role as a seed dispersal agent on Hong Kong hillsides; in Guangdong, however, it is sold as food in markets. Photo by Gary Ades.

Plate 30
Rhesus macaques, *Macaca mulatta*, are Hong Kong's only indigenous monkeys, but whether the specimens photographed here are descendants of the original animals, or of individuals that were released subsequently, is not known. The distinctive features of rhesus macaques are shown by this female: in particular, the short tail, and the grey colour of the forequarters blending into rust on the back and hindquarters. Photo by John Fellowes.

Plate 31
Long-tailed macaques, *Macaca fascicularis*, were first recorded in Hong Kong during the 1960s. They are now quite abundant and have hybridized with rhesus macaques. The distinctive features of the long-tailed macaque shown by this male include the tail length and the rather uniform grey coat. Photo by John Fellowes.

Plate 32
The lower Lam Tsuen River before channelization during the early 1980s. Note the dense growths of aquatic plants which almost cover the water surface; these provided habitat for a range of animals which were eliminated by channelization. Photo by David Dudgeon.

Plate 33
The lower Lam Tsuen River after channelization. Photo by David Dudgeon.

Plate 34
Most stream insects have an aquatic larval stage which has the ecological function of feeding, growth and development, followed by a winged terrestrial adult stage devoted to reproduction and dispersal. A winged adult mayfly (Ephemeroptera) is shown here. Photo by Chong Dee-hwa.

Plate 35
The winged terrestrial adult of a Hong Kong stonefly (Plecoptera: Perlidae). Photo by Chong Dee-hwa.

Plate 36
The minnow, *Parazacco spilurus*, is typical of pools in upland streams. The well-developed tubercules on the gill cover and lower jaw of this mature male are a sign of readiness for breeding. Photo by Chong Dee-hwa.

Plate 37
With its flattened body and modified fins, *Pseudogastromyzon myersi*, is well adapted to clinging to rocks in fast current where it grazes algae from stone surfaces. Photo by Chong Dee-hwa.

Plate 38
Liniparhomaloptera disparis feeds exclusively on algae and, like *Pseudogastromyzon myersi*, is confined to clean, fast-flowing hillstreams. Photo by Chong Dee-hwa.

Plate 39
The predatory loach *Noemacheilus fasciolatus* is common in hillstreams, and is easily recognized by the conspicuous bands on the body. Photo by Chong Dee-hwa.

Plate 40
Plecoglossus altivelis is a distant relative of the salmon and, like those fish, spawns in freshwater but spends part of its life in the sea. This specimen was found in Tai Ho Stream on North Lantau, which is the only known habitat of *Plecoglossus* in Hong Kong. Photo by Chong Dee-hwa.

Plate 41
Ferret badgers, *Melogale moschata*. Very little is known about the ecology of these mammals, although they seem to be widespread in Hong Kong. Photo by Gary Ades.

Plate 42
The short-legged toad, *Megophrys brachykolos*, provides a good example of crypsis. This species is found around forest streams and its colouration blends in with the leaf litter lying on the soil surface. Photo by Michael Lau.

Plate 43
The common Indian toad, *Bufo melanostictus*, is cryptic when sitting motionless on the ground. This is Hong Kong's commonest amphibian and an important predator upon insects and other invertebrates; it is found in a wide range of habitats. Photo by Chong Dee-hwa.

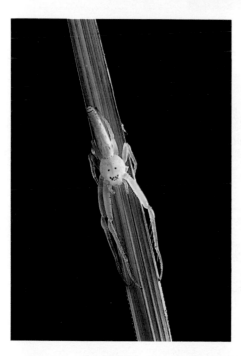

Plate 44
Crypsis is employed by this thomisid spider, *Oxytate hoshozuna*, as a means of evading its predators, but also to conceal itself from insect prey which can be grabbed should they venture too close. Hong Kong's spiders are exceptionally diverse but, to date, have been neglected by local biologists. Photo by Chong Dee-hwa.

Plate 45
Stick insects (Phasmatodea) are cryptic in terms of their colour and body form, and cannot be distinguished easily from the foliage (in this case mountain tallow, *Sapium discolor*) upon which they graze. Photo by Chong Dee-hwa.

Plate 46
The arboreal forest snail, *Cryptosoma imperata*, grazes plant foliage throughout the warm, wet summer months, but is inactive during the winter which it spends buried in the soil. This snail, which was first described in 1859, seems to be known from nowhere else but Hong Kong. Photo by Chong Dee-hwa.

Plate 47
The body of the Chinese pangolin, *Manis pentadactyla*, is covered with horny scales. Large claws on the front feet are used for excavating termite and ant nests, and the pangolin captures and eats these insects by using a long (25 cm), sticky tongue. Photo by Gary Ades.

Plate 48
The Chinese leopard cat, *Felis bengalensis*, resembles a large, domestic cat, but has a distinctive white blotch on the back of each ear. Like civets, it is a common market mammal in Guangdong. Photo by Gary Ades.

Plates 49 & 50
Two ways of being a skink: the brown forest skink (*Lygosoma indicum*; Plate 49) is a leaf-litter dweller and cryptic. By contrast, the five-striped blue-tailed skink (*Eumeces elegans*; Plate 50) inhabits hillside grassland, where its bright colours must make it extremely conspicuous. Skinks will shed their tails when attacked, and it is possible that the conspicuous blue tail distracts the predator allowing the skink to escape. Photos by Michael Lau.

Plate 51
Hong Kong's largest (up to 35 cm long) and most widespread skink is the Chinese skink, *Eumeces chinensis chinensis*. It particularly favours lowland areas and is likely to be a significant predator upon larger invertebrates. Photo by Chong Dee-hwa.

Plate 52
The crested tree lizard, *Calotes versicolor*, is largely arboreal and found mainly in shrubland. It can reach 40 cm in length and is incorrectly supposed to be poisonous. Photo by Chong Dee-hwa.

Plate 53
The Chinese cobra, *Naja naja atra*, is a common, widespread and extremely venomous snake. It feeds on amphibians, reptiles (including other snakes) and small mammals. Photo by Michael Lau.

Plate 54
The largest terrestrial predator (up to 6 m long) in Hong Kong, the Burmese python, *Python molurus bivittatus*, is protected by law. Photo by Michael Lau.

Plate 55
The Hong Kong cascade frog, *Amolops hongkongensis*, is known only from Hong Kong and is protected by law. Note the well-developed suckers on the toes which enable the frog to hold on to slippery rocks in fast currents and waterfalls. Photo by Michael Lau.

Plate 56
The water monitor, *Varanus salvator*, which can reach over 2 m in length, is probably now extinct in Hong Kong, although a few specimens obtained from local markets have been released here in recent years. Photo by Gary Ades.

Plate 57
Hong Kong's most celebrated — and perhaps most threatened — animal is Romer's tree frog, *Philautus romeri*, which (following the destruction of Chek Lap Kok) is known only from three of the Territory's islands. Photo by Michael Lau.

Plate 58
A paper wasp (*Vespa* sp.) nest; as with all social insects, the members of this colony are closely related to each other because they are offspring of the same female (the 'queen'). Photo by Chong Dee-hwa.

Plate 59
A *Popilla* chafer; in June, these beetles undertake nuptial flights but can be found feeding upon flowers. Photo by Chong Dee-hwa.

Plate 60
Hill fires are a common sight during the dry season, and are a major influence on Hong Kong vegetation. Photo by Gary Ades.

Plates 61 & 62
Seasonal change in the Hong Kong countryside: hillside grassland (Tai Mo Shan) during the wet season (Plate 61) and hillside grassland (Pat Sin Range) during the dry season (Plate 62). Photos by Michael Lau.

Plate 63
A marshy abandoned field in a upland valley (Pak Shek Kin) that was probably once used for rice cultivation. Such habitats can be important amphibian breeding sites. Photo by Michael Lau.

Plate 64
Tung Chung Stream on North Lantau is important as a habitat for rare fishes, and is one of the few relatively large streams in the Territory to have escaped the worst effects of pollution. Photo by David Dudgeon.

Plate 65
Tai Po Kau Forest Stream. Photo by Chong Dee-hwa.

Plate 66
A plantation in Shing Mun Country Park made up entirely of Australian paperbark, *Melaleuca quinquenervia*. Photo by Chong Dee-hwa.

Plate 67
The Javan mongoose, *Herpestes javanicus*, is a rare nocturnal carnivore, but breeding in Hong Kong has been confirmed recently. Photo by Michael Lau.

Plate 68
The lesser sulphur-crested cockatoo (*Cacatua sulphurea*) originates in Indonesia but has been introduced into Hong Kong. Flocks of these birds are a conspicuous and noisy element of the avifauna on northern Hong Kong Island, where pairs have been recorded as nesting in holes in large trees. Photo by Chong Dee-hwa.

4

Seasonality

Life-cycle events and fluctuations in the abundance of temperate-zone animals and plants are often seasonal; spring-time breeding, mammalian hibernation, migration by birds, and loss of leaves by deciduous plants during the autumn are familiar examples. Periodic phenomena are also known in the tropical flora and fauna, even in equatorial regions where rainfall and temperature are equable throughout the year. In the seasonal tropics, changes in animal abundance or behaviour, and the timing of leafing, flowering and fruiting by plants, may reflect rainfall patterns, but other factors such as light intensity, phases of the moon or photoperiod may play a role. Interspecific interactions might also influence seasonality where, for example, there is a need for flowering to coincide with the main period of activity of pollinators. The range of influencing factors is such that it is difficult to make general statements about the seasonality of tropical plants and animals, whether terrestrial or aquatic, since the causes of seasonality may vary from one species to the next. Indeed, the factors determining seasonality and population dynamics in the majority of tropical organisms remain a matter for speculation.

Temperature plays a major role in the ecology of most organisms, but its influence is modified by responses to other factors which do not remain constant and may vary with the same periodicity as temperature. For example, when we think of changes in the abundance of insects in relation to the seasonality of Hong Kong's climate, it is difficult to assess the relative influence of fluctuations in temperature, rainfall and day length because increased rainfall during the summer is accompanied by a rise in temperatures and greater day length. Further complications arise if the same factor can influence terrestrial and aquatic species differently. For example, the tropical dry season is an adverse period for many terrestrial plants and animals but, unless streams, reservoirs or marshes dry out, the effects of a dry spell on freshwater organisms are difficult to determine. However, in certain situations, along the gently-shelving shores of reservoirs such as Plover Cove, for example, mass stranding of slow-moving animals and aquatic plants can occur as water levels retreat during the dry season. Because of the potential for differences in the importance of particular climatic factors on land and in water, the seasonality of animals in these two environments will be considered separately. The majority of plants are, however, terrestrial and their seasonality will be considered first here.

Plant seasonality: leafing

In areas with an extreme cold or dry season, many herbaceous plants adapt by completing their life cycles in less than one year and passing the unfavourable season as resistant seeds. Winter in Hong Kong, however, is neither very dry nor very cold and annual plants are unimportant in the hillside flora. Although most herbaceous species die back to ground level in winter, their below-ground parts survive and sprout in spring.

Woody plants also synchronise their vegetative growth with climatic seasonality. Probably all aspects of growth are seasonal to some extent but the most visible and significant phenological events concern the leaves. The layman's classification of woody plants as evergreen or deciduous seems so obvious and fundamental that it is hard to believe that, in the tropics, these are just extremes of a continuum of leafing behaviour. The majority of trees and shrubs in Hong Kong have leaves that last about one year, and a major flush of leaf production in February, March or April. If the previous year's leaves senesce and fall before the new ones expand, the plant is deciduous. If the leaf generations overlap, it is evergreen. For most species in Hong Kong, this period of overlap is short, and the plants can be termed 'leaf-exchanging'.

In several common tree species, different individual plants show a range of leafing behaviour and timing. In *Acronychia pedunculata*,

Diospyros morrisiana and *Reevesia thyrsoidea*, for instance, adjacent individuals can be found in March completely leafless, exchanging leaves, and with a complete set of old leaves. Most individuals of *Cratoxylum cochinchinense* are leafless for a month or more but some, at wetter sites, are either only briefly deciduous or evergreen leaf-exchangers. Several other 'deciduous' species also have a few, more-or-less evergreen individuals. In fact, obligate deciduous plants are not common in the local tree and shrub flora. All trees of *Sapium discolor* are leafless from December to March and a few other widespread species, such as *Celtis sinensis*, *Rhus succedanea* and *Ilex asprella*, are consistently leafless for more than a month. A number of less common species found in a variety of habitats are also deciduous, but the overall aspect of most woody vegetation in Hong Kong is evergreen.

Most trees and shrubs complete their major leaf exchange by April, although many have one or more additional leaf flushes later and some fig species change their leaves completely more than once a year. However, a few tree species, including *Artocarpus hypargyreus* and *Garcinia oblongifolia*, do not exchange leaves until May, while *Sterculia lanceolata* and *Ormosia emarginata* change leaves in mid-summer. *Sterculia* trees are often briefly, and incongruously, leafless at this time. Perhaps significantly, these plants are all members of tropical genera which, in Hong Kong, are near the northern limits of their distribution. All of the major phenological events in these species, including leafing, flowering and fruiting, are compressed into the hot, wet 'tropical' months of the year.

The example of *Sterculia* suggests that the explanation for deciduousness in Hong Kong is not straightforward. Temperatures here rarely fall low enough to limit physiological activity and it is not obvious why winter water shortage should affect some species so much more than others. Moreover, the timing of both leaf fall and subsequent re-leafing seems to be unaffected by year-to-year variation in the amount and timing of rainfall. *Sapium discolor* is briefly deciduous even in the almost aseasonal climate of Malaysia, suggesting that this behaviour is an inherent property of the species, exaggerated and synchronized by Hong Kong's more seasonal climate, but not necessarily evolved in response to it. It may be significant that *Sapium* bears ripe fruit at the time when its leaves are turning red prior to being shed. The fruits are inconspicuous but the whole tree can be identified from more than a kilometre away by the colour of the senescing leaves. A similar advantage of early leaf senescence in attracting frugivores and seed dispersers can also be suggested for *Rhus succedanea*, and probably other species.

The timing of leaf flushing in many species may have evolved to ensure that the vulnerable young leaves expand and mature while insect activity is still low in early spring. Although no quantitative study has been made, leaves expanded in February and March do seem to receive less damage from herbivores than leaves expanded in April and May, and the peak in herbivore grazing activity seems to occur in May and June (see Chapter 7: Herbivores). Early flushing as an escape from herbivores is possible for woody plants because their deep roots give them access to sufficient water at a time when invertebrate populations are, apparently, limited by seasonal drought, low humidity and perhaps temperature.

Plant seasonality: reproduction

The timing of flowering and of fruiting in a particular species are obviously not independent: the time taken for a ripe fruit to develop after pollination is more-or-less fixed. However, in evolutionary time, this link seems surprisingly flexible and related species can have very different fruit-development times. All seven of the common species of holly (*Ilex*) in Hong Kong flower between March and May. Six of them fruit the next winter, after a fruit-development period of 220 to 280 days. The other species, *Ilex asprella*, in contrast, ripens its fruits in June, only 70 days after flowering. Even more striking are the leguminous climbers, *Dalbergia benthami* and *D. hancei*, which take 51 and 300 days, respectively, to ripen their very similar, winged fruit. Other genera with a wide range of fruit-development times include *Rhus* (two species, 65 and 163 days), *Smilax* (three species, 81 to 224 days), and *Zanthoxylum* (three species, 132 to 280 days). Looking at the woody shrubland flora as a whole, the extremes recorded so far are 45 days (*Lasianthus chinensis*) and 365 days (*Eurya nitida*).

These data suggest that, over evolutionary time, a species can adjust its flowering and fruiting times more or less independently so both are near the optimum for that species. But what determines this optimum? Opportunities for successful pollination and seed dispersal would seem the obvious answer but it is not that simple. Both flowering and fruit development need energy and nutrients, and the availability of these could constrain reproductive phenology. Flowering is often tightly linked with vegetative growth, occurring at a fixed point in relation to the annual growth cycle. Fruiting and successful seed dispersal are useless if conditions are unsuitable for germination at that time and if the seed cannot survive until they become suitable.

The factors which determine the success of pollination and seed dispersal will be different for each species, so the optimum reproductive phenologies are also likely to differ among species. We do know, however, that there are striking and predictable seasonal patterns in reproductive phenology when the hillside plant communities are studied as a whole. Most work has been done on woody species. For these, community patterns of both flowering and fruiting are highly seasonal with little variation between years (Fig. 3). There is a flowering maximum in May, when about a third of shrubland species are in flower, and a minimum in December, when, of the common species,

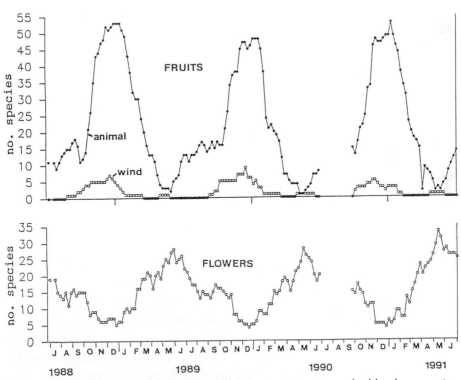

Fig. 3 Seasonal patterns of flowering and fruiting in a Hong Kong shrubland community. Weekly records of the number of plant species in fruit (upper: animal- and wind-dispersed species separately) and in flower (lower) along a 5.8 km route in Pokfulam Country Park: July 1988 to June 1991.

only *Gordonia axillaris* and *Schefflera octophylla* are in full flower. The community fruiting phenology is virtually a mirror image: a maximum in late December and early January, with about half the species in fruit, and a minimum in May, when no common species is at its best. The fruiting maximum involves more species than the flowering maximum largely because most species bear ripe fruit for longer than they bear flowers.

It is not difficult to come up with plausible explanations for these patterns, but it may be impossible to test such hypotheses. The May flowering maximum might reflect a peak in the abundance and activity of potential pollinators after a winter low. Some pollinators, such as carpenter bees (*Xylocopa* spp.), are active only in the summer months and most are rarer in winter. The flowering maximum may also be explained, at least in part, by the fact that flower production tends to follow re-leafing in many species. Interestingly, the grasses, which are the only major group of hillside flowering plants that are pollinated by wind instead of insects, tend to flower in late summer.

The explanation for the mid-winter fruiting maximum seems more obvious. Not only is this the time when migrants swell the density of fruit-eating (and seed-dispersing) shrubland birds to a maximum, but it also coincides with a minimum in insect activity so that resident insectivore-frugivores, such as the bulbuls (*Pycnonotus* spp.) and white-eye (*Zosterops japonicus*), switch to an almost entirely frugivorous diet. Thus we can view fruiting seasonality as having evolved to take advantage of an increase in the number of frugivorous and seed-dispersing birds in winter. Unfortunately, there is one major flaw in this argument: wind-dispersed shrubland species exhibit a fruiting maximum at the same time! This suggests that there is some other advantage in mid-winter fruiting, perhaps to do with having seeds in place for an early start with the first rains. An alternative (and non-exclusive) explanation is that wind dispersal may also be more effective in winter because a dry seed can be carried further or because sparser foliage increases wind speeds near ground level.

Species that flower before the May peak tend to have longer development times than those that flower after the peak. Two extreme patterns can be recognized: species that flower one winter and fruit the next (e.g., *Eurya* spp., *Gordonia axillaris* and *Rhododendron* spp.) and those that both flower and complete fruit development in one summer (e.g., *Artocarpus hypargyreus*, *Garcinia oblongifolia*, *Gnetum montanum* and *Morinda umbellata*). No species flowers one summer and fruits the next and only one species, *Schefflera octophylla*, flowers and fruits during the same winter. A few species, such as *Lasianthus chinensis*, flower and fruit twice a year while others have prolonged flowering and fruiting seasons. The extreme case is the common shrub, *Breynia fruticosa*, which flowers for the ten warmer months of the year, although it fruits rather erratically.

One consequence of the community pattern of reproductive phenology is that the majority of species bear unripe fruit throughout the hottest and wettest time of the year, when the threat of attack by insects or microbial damage would presumably be greatest. The risks of a relatively long fruit-development time must be outweighed by the evolutionary advantages of the observed flowering and fruiting times. Surprisingly, mean fruit-development times in Hong Kong are longer than in cooler climates to the north, where the fruiting peak is in late summer or autumn.

Bird migration

It is likely that all animals in Hong Kong show some correlation between life-cycle events and climatic seasonality. In the majority of species, the number of individuals changes through the course of the year, either through movements of animals in and out of the Territory (migration), or through seasonality of reproduction, or both.

The migration of birds is perhaps the most striking example of seasonality in Hong Kong. Less than a quarter of the bird species recorded from the Territory are present year round and, even among these 'resident' species, there are often many individuals that migrate. The majority of Hong Kong's birds are either winter visitors, overwintering in Hong Kong but breeding to the north and northeast (northern China, Mongolia, Japan, Korea and Siberia), or passage migrants which pass through Hong Kong on their way to and from wintering grounds nearer the equator. Unfortunately, the migratory routes of the majority of non-coastal birds passing through Hong Kong are uncertain. Most of the migrants are Palaearctic species (perhaps better termed 'Palaeotropical' as they spend most of the year in the tropics) but the winter influx also includes additions to the resident populations of Oriental species, such as the Chinese bulbul (*Pycnonotus sinensis*) and white-eye (*Zosterops japonicus*), and additional Oriental species, such as the black bulbul (*Hypsipetes madagascariensis*). Palaearctic migrants dominate the bird fauna in winter, in terms of species if not individuals.

Bird migratory movements are probably driven largely by the availability of food. In general, the insectivorous migrants arrive first (September and October) and the frugivores and partial frugivores last (November and December). This is the order in which we would expect food availability to decline at higher latitudes, with insect numbers falling first as the weather cools and fruits persisting on trees until consumed. The situation is more complicated with fruits, however, since frugivory is often a mutualistic relationship involving seed dispersal, and the timing of fruit ripening is also subject to natural selection.

The fruiting peak in Hong Kong is in winter and a high proportion of overwintering birds eat at least some fruit. Probably all of the many species of migrant robins (such as *Luscinia* spp.) and thrushes (*Turdus* spp.), for instance, eat fruit while in Hong Kong, although its contribution to the total diet varies tremendously. Insects, in contrast, are harder to find in winter, so it is not surprising that many strict insectivores merely pass through Hong Kong *en route* to less-seasonal areas nearer the equator. However, several strict insectivores, such as the yellow-browed warbler (*Phylloscopus inornatus*), do overwinter in Hong Kong. Changes in the supply of seeds for granivores have not been investigated locally, but the abundance of overwintering buntings (*Emberiza* spp.), which seem to be largely granivorous in Hong Kong, suggests that it is always adequate.

By May, both winter visitors and passage migrants have flown north or northeast to take advantage of the seasonal abundance of insects on their summer breeding grounds. Few species are summer visitors to Hong Kong, but some of these are common and probably ecologically important. The most conspicuous are aerial insectivores: the swallow (*Hirundo rustica*) and the black and hair-crested drongos (*Dicrurus macrocerus* and *D. hottentottus*). All three are seen in small numbers in winter but are far more common in summer. Two other common summer visitors, the Indian and plaintive cuckoos (*Cuculus micropterus* and *Cacomantis merulinus*), are more often heard than seen. Both species are brood-parasites, laying their eggs in the nests of other species. The Indian cuckoo parasitises a (mainly) summer visitor, the black drongo, while the plaintive cuckoo parasitizes the resident tailor-bird (*Orthotomus sutorius*). The resident koel (*Eudynamys scolopacea*) also parasitizes resident birds, such as black-necked starlings (*Sturnus nigricollis*) and crested mynahs (*Acridotheres cristatellus*).

Seasonality and breeding by terrestrial animals: vertebrates

Most birds in Hong Kong, from laughing-thrushes to magpies and swallows to white-eyes, breed in spring and early summer (March to July) although nesting continues throughout the summer in some species. Those birds with a long breeding season produce multiple broods: probably two in the case of the crested mynah and great tit (*Parus major*), up to three in the black-necked starling and yellow-bellied prinia (*Prinia flaviventris*), and even more in the white-eye and tree sparrow (*Passer montanus*). However, the nuptial flights of black kites (*Milvus migrans*) are a common sight throughout the winter and breeding activity of some other birds is initiated early too. Great tits begin nesting in February and, later that month, the calls of greater coucals (*Centropus sinensis*) begin to resound around the Territory and continue throughout spring. Koels and Indian cuckoos are heard soon after, in late February or the first days of March.

Moulting (shedding and replacing the feathers) is another important seasonal activity in birds. In general, resident birds moult after the end of the breeding period in Hong Kong but before the coldest months of winter. Moulting must be a major drain on energy as well as reducing flight efficiency, so it is clearly incompatible with collecting food for hungry young birds.

Reproductive activity of many birds coincides with an increase in the abundance and diversity of insects which the adults need to feed their nestlings. Breeding by insectivorous bats, such as the bent-winged (*Miniopterus* spp.), roundleaf (*Hipposideros* spp.) and horseshoe bats (*Rhinolophus* spp.), also takes place in spring, although mating in *Rhinolophus* and *Miniopterus* occurs in late autumn of the previous year. In the latter, the development of the foetus is delayed until spring (because implantation is delayed), while in the former genus (and all insectivorous bats except *Miniopterus*) sperm are stored for several weeks prior to fertilization of the egg. Females give birth to a single offspring in communal nursery colonies, from which most males are excluded, and these nurseries may be the same sites (usually caves or abandoned mines) which the bats use to overwinter. Although nursing mothers of some species carry their new-born offspring with them during flight (e.g., the Japanese pipistrelle, *Pipistrellus abramus*), young of other bats (such as the bicoloured roundleaf bat, *Hipposideros pomona*) are left suspended from the walls or roof close to the entrance of the nursery site when the females leave the colony to forage at night. Upon return, females go to the spot where their offspring were left several hours before, retrieve them and carry the offspring to the nursery area where they are suckled. Within four weeks of birth the young bats are flying, and they are making their own foraging trips to catch insects by about eight weeks old. The females do not give birth again until the next year. Only in the Japanese pipistrelle is there any evidence of production of a second brood within a single year.

In winter, some insectivorous bats (particularly horseshoe bats) enter a kind of torpor or dormant state, which may be a means of conserving energy during the months when insects are in short supply. Breeding by these insectivores coincides with increased prey abundance during spring and early summer, but it is notable that reproduction of fruit bats (*Cynopterus sphinx* and *Rousettus leschenaulti*) takes place at the same time, during the community-level minimum in fruit availability. Only figs are abundant at this time and these are not good sources of protein. Some fig species, however, bear huge crops, and ease of foraging may compensate for the relatively low nutritional value. Fruit bats elsewhere in Asia are known to eat young leaves and shoots when fruit are in short supply but we have no evidence of this for Hong Kong.

Little is known about reproductive seasonality in other Hong Kong mammals, although the presence of juveniles suggests that breeding by the introduced Pallas's squirrel (*Callosciurus erythraeus*) takes place during much of the year. Up to three young are produced at a time. Sexually-active males are readily identified by their characteristic rust-red pelage on the abdomen and flanks, and the fact that mating has been observed in December suggests that temperature does not limit breeding activity. Ball-shaped nests of twigs and leaves can also be seen also throughout the year, but the squirrels probably do not use the nests exclusively for breeding. Hillside rats (*Niviventer fulvescens* and *Rattus sikkimensis*) also appear to reproduce throughout the year, although there may be seasonal variations in breeding intensity. *Rattus rattus flavipectus*, which is more common in lowland grassy areas, likewise seems to reproduce in most months of the year. Two other rats, *Rattus rattus rattus* (the black rat) and *Rattus norvegicus* (the brown rat), occur only in the vicinity of human habitation, and breed all year round. Unfortunately, nothing is known of the habits of Hong Kong's largest and scarcest rat, the greater bandicoot rat (*Bandicota indica*).

One interesting aspect of mammalian seasonality in Hong Kong concerns the South China tiger (*Panthera tigris amoyensis*) and the leopard (*Panthera pardus*). Although now extinct locally, and highly endangered throughout their range in China, there are records of big-cat predation upon domestic cattle in Hong Kong during the 1930s. Apparently, these carnivores were winter visitors to the Territory, and may have been driven south by cold or a shortage of food. Further investigation of this point is impossible: neither species has been sighted here for over 40 years.

Terrestrial snakes and lizards in Hong Kong show a clear pattern of seasonality. Most snakes and lizards hibernate in winter, but geckoes and snakes will emerge briefly during warm spells when the temperature exceeds 16°C. The Burmese python (*Python molurus bivittatus*), Hong Kong's biggest snake, does not hibernate, but fasts during the cooler months (November to February). The majority of the 41 species of non-marine snake found locally breed between April and September, although young are sometimes produced during February and March. Six species are live-bearing; they bear young directly rather than laying an egg which must be incubated before hatching. Springtime initiation of breeding seems to be the rule for lizards in Hong Kong, but egg laying may continue throughout the summer.

Seasonality and breeding by terrestrial animals: invertebrates

Terrestrial invertebrate (particulary insect) seasonality is striking in Hong Kong, and the cool, dry winter months seem to be inimical to many species. For example, the arboreal snail *Cryptosoma imperator* (Ariophantidae) buries itself in the soil and becomes inactive during the winter — a behaviour which is typical of most Hong Kong land snails. While low temperatures can restrict invertebrate activity, dryness is also an important limiting factor. Low humidity and a lack of free water for drinking must contribute to the lack of insect activity during much of the autumn and winter when temperatures are mild but the soil surface, leaf litter and much of the standing vegetation are tinder dry. Although few insects are active as adults during the winter, feeding and growth by larvae of many species continues, whether they are burrowing in the soil, concealed among leaf litter, or boring through living and dead wood and other plant tissues. Other species spend the winter in a resting or diapause stage, usually as an egg or pupa, although some insects can overwinter as adults. The annual pattern of insect abundance is thus marked by a depression during the autumn and winter, with a rapid rise in April to a peak in May, followed by a slight decline and a second peak in late July. By October, abundance and species richness starts to tail off and it falls to a low level in November. The timing of the autumn decline depends upon temperature and rainfall, and it may be delayed by a particularly hot and wet autumn.

The mating song of the yellow-spotted black cicada (*Gaeana maculata*), which is first heard at the end of March or early in April, marks the emergence of adults from the soil where their larvae have spent much of the year, and is a distinctive herald of things to come. A host of other species are soon abroad. *Gaeana maculata* is soon replaced by another cicada, the large brown *Cryptotympana mandarina*, from May through July, and a diverse array of insects, such as, butterflies, grasshoppers, beetles, etc., are present throughout the summer and early autumn, although others have a more confined period of adult life. In April, May and early June, chafers (related species in North America are known as June bugs) are frequent nocturnal visitors to street lamps or lights in buildings. These beetles spend much of the year in soil, where the grub-like larvae feed on roots. The brown *Holotrichia* adults are the earliest and most frequently encountered chafers (Scarabeidae: Melolonthinae) in Hong Kong, and the larvae can be significant crop pests. The metallic-green *Popilla* chafers (Scarabeidae: Rutelinae) have similar habits and, in early June, adults may be found in flowers which they eat, petals and all.

We do not know why some terrestrial insects persist throughout the summer and may pass through several generations within this favourable period, while others such as chafers, cicadas, and the spectacular lantern bug (*Pyrops candelaria*: Fulgoridae) pass through only a single generation each year. Final body size must provide part of the answer; bigger bodies need a longer larval feeding time, and it is noticeable that insects subsisting upon food which is low in nutrients, such as wood, take at least a year to complete their life cycles. Although it is not only the larger insects which appear only briefly, it is curious that many relatively large insects such as chafers, cicadas and lantern bugs (among others) have brief adult lives when, for example, paper wasps (Vespidae), which are approximately the same size as chafers, are common throughout summer and most of the autumn. Perhaps the explanation for the short adult lives of some larger insects lies in the fact that longer-lived adults would gain no extra benefit from a prolonged life, and thus there has been no natural selection for longevity. This could, in part, be due to intense predation pressure upon large insects, both from birds and bats. Indeed, chafers form a major part of the diet of certain bats for a short period in the year, reflecting the fact that they are clumsy, noisy fliers which are easy to detect and capture. In an environment where the risk from predators is high, evolutionary success (or fitness, as described in Chapter 1) will depend on the ability to find a mate and complete reproduction as quickly as possible; short-lived, efficient 'breeding machines' are the result because no additional evolutionary advantage is gained by a prolonged adult life after breeding. In other words, those long-lived animals which persist after breeding leave no more offspring, and pass on no more genes, than insects which die immediately after the last eggs are fertilized or laid.

Why, then, do paper wasps have a prolonged adult life? In this case, the underlying causes are complex. Common Hong Kong paper wasps, such as species in the genera *Vespa* and *Polistes*, are social insects which live in a colony of related individuals. The entire colony is descended from a single queen wasp that is fertilized by one (or a very few) males during a nuptial flight in the autumn. The fertilized queens overwinter as torpid adults concealed in cracks and crevices, and emerge in spring to build a small paper-like nest comprising a mixture of wood shavings and saliva. A small number of eggs and larvae are produced and tended by the queen but, once they emerge from their pupae, the new recruits become the colony's workers with duties which include caring for the queen and extending the nest. They also attend to their many siblings produced by the queen, who becomes little more than an egg-laying machine as the worker population increases. All members of the colony at this stage are female. In social wasps and their relatives (the ants and bees which, together with the wasps, make up the Hymenoptera) unfertilized eggs develop into males and fertilized eggs into females. By controlling the release of sperm which were stored in her body after mating during the nuptial flight, the queen is able to manipulate the sex ratio of the colony and produce eggs that will develop into female workers.

While the ratio of workers to eggs and larvae is low early in the life of the colony, the workers cannot deliver sufficient food to their larval sibs and the resulting adults are undersized and do not breed. As the worker population increases during the summer, more food can be brought into the nest and the well-fed larvae pupate into sexually mature wasps — both males (which do not work) and future queens. These wasps leave the nest to mate and found new colonies after over-wintering; so too does the original queen if she is not senile. The workers remain with the nest for some time but the demise of the colony is inevitable in the absence of a queen.

The special feature of the paper wasp colony is that it is eusocial, consisting of a group of individuals with the same parentage and hence a similar genetic composition. Here, some individuals do not reproduce at all. As is apparent from Chapter 1, if individuals sacrifice their own reproduction, then the genes for this sacrifice will be lost unless these individuals aid the reproduction of others who share the genes. The evolutionary success of non-breeding (altruistic) paper wasps depends upon ensuring that a large number of their sisters (which carry most of the same genes) survive to become future queens and colony founders. The altruistic individuals increase their inclusive fitness (determined by the reproductive success of an individual plus that of relatives which share the same genes by common descent) by tending their sisters, and their success in this respect is increased if they are long lived. Significantly, altruism of this type (and eusocial behaviour) is confined to ants, certain bees and wasps, and termites, i.e., species with high relatedness among colony members because of their descent from a single queen. Under these circumstances, the death of an individual honey bee (*Apis mellifera*), for example, following the use of its sting to defend the colony has little evolutionary significance for that individual as long as the survival and reproduction of some sisters of the bee are ensured.

The example of paper wasps demonstrates that the seasonality and adult flight period of certain Hong Kong insects depends upon their evolutionary history, in this case, upon the evolution of social behaviour. Species with rapid development and short adult lives (some butterflies and many small herbivorous insects) are the outcome of natural selection for rapid growth and the ability to complete several generations within the warm, humid summer months. Under these circumstances, evolutionary success accrues to individuals with the fastest rate of production of viable offspring. This strategy is probably typical of most small terrestrial invertebrates in Hong Kong and, in such species, seasonality is imposed by low temperatures or dryness limiting breeding during the winter. This gives rise to patterns of abundance typified by invertebrates such as woodlice (isopod crustaceans) which are found among leaf litter in Hong Kong forests where they can reach densities of 400 individuals/m^2. Of these, *Burmoniscus ocellatus* and *Formosillo raffaelei* population densities peak at the end of summer (September and October) but are low in winter (December through March), increasing again during the breeding period which begins in April and continues until October. Similarly, species richness, abundance and flight activity of necrophagous flies (especially the family Calliphoridae), whose larvae feed on carrion (see Chapter 7), increase in the summer and decline between December and March. That this pattern parallels the seasonal trends seen in woodlice (a group of terrestrial crustaceans which are both ecologically and phylogenetically unrelated to carrion flies) emphasizes the generality of the summer-maximum winter-minimum change in the density of terrestrial invertebrate populations.

What causes this pattern? Statistical analysis of the relationship between woodlice or carrion fly abundance and climatic factors (rainfall, temperature, relative humidity) shows that temperature is closely related to population size; the relationship with rainfall is similar but less statistically significant. Such data do not demonstrate unequivocally that temperature is the main factor driving seasonality of many terrestrial animals (because parameters such as day length will co-vary with temperature) but this finding makes biological sense given our knowledge of the effects of temperature on metabolic processes which will, in turn, influence rates of development and population increase. This explanation is, nevertheless, unsatisfactory because winter temperatures in Hong Kong are in the same range as summer temperatures in parts of the temperate zone, where invertebrates have adapted to them successfully. In other words, it is evident that invertebrates could continue to be active during the cooler months in Hong Kong if there were no other limiting factors. In addition to temperature, humidity and rainfall affect animal populations, and low humidity and lack of free water in the dry season will restrict the activity of many invertebrates, especially small species with a high surface-to-volume ratio which are prone to lose water through the body surface. It is of interest that habitat use by mammals is influenced also, and hillside rats seem to change their home ranges during dry periods so as to include a perennial stream or other source of drinking water. There is also likely to be an additional indirect effect of climate upon most animals through its influence on food plants. In the same indirect manner, seasonality of breeding by mammals and birds may be induced by rising temperature if the availability of insect food is enhanced in warmer weather.

Despite their seductiveness, all-inclusive explanations for animal seasonality in Hong Kong should be treated with caution. For example, certain scarabeid dung beetles which lay eggs inside dung pads where their larvae will feed and grow, breed only during the drier months because heavy rain waterlogs the dung. By contrast, swarms of termites such as *Macrotermes barneyi* and *Coptotermes formosanus* undertake nuptial flights which coincide with summer rains. These flights occur sporadically during the warmer months when vast numbers of individuals leave termite colonies to search for mates. Initiation of swarming is highly synchronized among colonies, and seems to be correlated with sudden drops in atmospheric pressure associated with the onset of heavy rain. In this case, temperature could be viewed as a priming mechanism for termite reproduction, while factors such as pressure or rainfall initiate synchronized flight and ensure that many potential mates are abroad at the same time.

As a broad generalization, animal life cycles and seasonality in terrestrial habitats in Hong Kong are driven by the combined effects of temperature, rainfall and humidity. Seasonal changes in day length may play a role also, but there is no evidence to support or refute this hypothesis. The same combination of factors will affect vegetation also and so indirectly influence food supply to herbivores. The spring increase in insect abundance stimulates breeding by birds, bats and perhaps other mammals, so that biological activity in the terrestrial environment is heightened greatly during the early summer. However, the precise timing and duration of breeding activities in an individual species will depend on the interaction of a variety of factors that are particular to that species, such as mating system and social behaviour, diet, evolutionary history or phylogeny, and rates of growth and development. As regards insects, one thing is clear: while there may be some consistency in the timing of life-cycle events (i.e., there is qualitative consistency), the details will vary from year to year with an individual species becoming abundant in one year but scarce for one or more of the succeeding years (i.e., there is quantitative inconsistency). In other words, while adults of the conspicuously-spotted cicada *Gaeana maculata* emerge in late March of every year (a qualitative observation), their abundance varies among years (quantitative inconsistency). Such fluctuations cannot be explained readily, but could reflect year-to-year variations in the weather which will affect conditions for insect growth and development. Alternatively, or additionally, fluctuations in the abundance of a particular insect could result from changes in the populations of the predators and parasites which limit insect numbers. A temporary escape from limitation allows the build-up of dense insect populations until such a time that the numbers of natural enemies recover, or weather conditions change and the environment becomes unsuitable so precipitating a population crash.

Seasonality of aquatic animals: invertebrates

We know more about the seasonality of aquatic animals in Hong Kong than we know about terrestrial animals (except birds), although our knowledge of the land fauna is increasing. In Hong Kong streams, summer spates or floods associated with monsoonal rains (see Chapter 2) can scour the stream bed so affecting the entire community, and the wet season may be an inimical time for some plants and animals in these habitats. In small streams, flow volume may decline dramatically during the dry winter months, and this may be a stressful period for rheophilic ('current-loving') species even if flow does not cease. In addition, given that seasonal changes in air temperature will affect Hong Kong freshwaters, they are likely to influence the population dynamics of freshwater animals.

Data on the periodicity of reproduction in freshwater molluscs suggest that this is the case for some species at least, although responses to fluctuating temperature and rainfall are not consistent within the group. Tiny bivalves of the family Sphaeriidae (*Musculium lacustre*, *Pisidium clarkeanum* and *P. annandalei*), which are found in slow-flowing lowland streams and drainage channels, breed throughout the year and do not seem to be limited by temperature. Seasonality is imposed by stream discharge, and washout during summer spates causes a decline in abundance. Other freshwater bivalves, such as the freshwater mussel (*Limnoperna fortunei*), breed twice, in summer and again in November or December, whereas the sand clam (*Corbicula fluminea*), reproduces in spring and autumn. Another pattern is seen in the swan mussel (*Anodonta woodiana*)(Fig. 4), which breeds only once each year, in early summer (May–July), and has larvae (termed glochidia) that are parasitic upon the fins and gills of fishes. Clearly, there is no uniformity in breeding seasonality among freshwater bivalves in Hong Kong, and the picture is complicated further when other molluscs are considered.

Among temperate pulmonate (air-breathing) snails, egg-laying in spring is stimulated by a rise in water temperature above a critical threshold and breeding ceases when temperature falls below that level in autumn. The warm season lengthens towards lower latitudes so lengthening the breeding season. In consequence, snails able to produce one generation per year in the northern part of their range, can produce two or more in warmer southern climates. Year-round breeding may be possible if temperatures remain sufficiently warm throughout the

Fig. 4 The swan mussel, *Anodonta woodiana* (Unionidae), breeds in early summer; larvae are parasitic upon the gills and fins of fish hosts for 10–14 days before taking up life as free-living filter-feeding juveniles. Only the inner and outer surfaces of the adult shell (length 8 cm) are shown here. Drawing by David Dudgeon.

year, in which case population densities depend upon the timing of droughts and floods associated with the dry and wet seasons. Reproduction by aquatic pulmonate snails in Hong Kong is reduced during the winter months, but egg-laying by common species (*Hippeutis cantonensis*: Planorbidae, *Physella acuta*: Physidae, *Radix plicatulus* and *Austropeplea ollula*: Lymnaeidae) (Fig. 5) can occur, albeit at a reduced rate, at water temperatures a low as 12°C. Indeed, high snail densities (up to 20 000 individuals/m²) in the middle and lower course of the Lam Tsuen River during the dry season indicate that cool winter temperatures do not restrict population growth under otherwise favourable circumstances, although populations crash after spates caused by monsoonal rains.

Among prosobranch freshwater snails, which breathe dissolved oxygen, most information for Hong Kong species concerns the family Thiaridae. The eggs and larval stages of these are retained inside a brood pouch on the head, and the young are released in the form of a crawling hatchling which is a perfect miniature of the adult. One species, *Melanoides tuberculata* (Fig. 5B), carries fully-developed hatchlings in the brood pouch throughout the year, and seems to have the potential to breed all year round. However, the hatchlings are released in response to rising temperatures, and so juveniles are most abundant in July and August. A similar pattern is seen in another

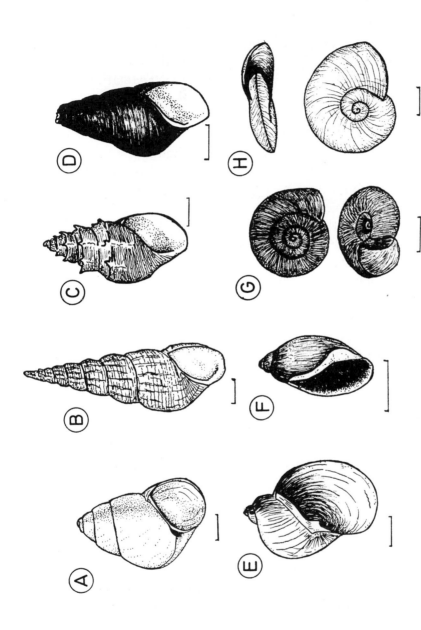

Fig. 5 Common freshwater snails of Hong Kong streams: A, *Sinotaia quadrata*; B, *Melanoides tuberculata*; C, *Thiara scabra*; D, *Brotia hainanensis*; E, *Radix plicatulus*; F, *Physella acuta*; G, *Biomphalaria straminea*; F, *Hippeutis cantonensis*. Scale lines = 5 mm (A, B, E & F), 4 mm (G), 3 mm (C), 2 mm (H) or 10 mm (D). Drawings by David Dudgeon.

prosobranch snail, *Sinotaia quadrata* (Fig. 5A), although this is not a member of the Thiaridae. In contrast to *Melanoides*, breeding activity of *Thiara scabra* (Fig. 5C) takes place during the coldest months in December and January. *Brotia hainanensis* (Fig. 5D) is the only snail that is common in stony hillstreams and the upper course of rivers in Hong Kong. Unlike other freshwater snails in Hong Kong, they have two breeding periods, one preceding and a second immediately following the summer monsoon. Such timing may be an adaptation to reduce juvenile mortality during spates.

The widespread occurrence of thiarids in Hong Kong streams could be a result of the evolution of a reproductive adaptation that involves brooding eggs and embryos until a crawling, miniature adult snail can be released from a brood pouch on the head. The ancestral thiarid was a marine snail and would have had planktonic larvae. Significantly, *Stenomelania*, the fourth genus of Hong Kong thiarid, is confined to freshwater creeks and marshes just above the high tide level and does not penetrate far up streams or rivers. These snails are similar to *Melanoides* in appearance and habit, yet they produce planktonic larvae from eggs which hatch in the brood pouch. The turbulence of hillstreams and the downstream loss of larvae would seem to have confined this genus to lowland areas, and release of well-developed crawling young during months when spates are unlikely to occur could account for the success of snails such as *Brotia* in running waters.

As is the case for bivalves, there is no evidence of a consistent influence of temperature upon the population dynamics of Hong Kong gastropods, and the winter depression in breeding activity seen in pulmonates is not translated into a decline in abundance in habitats such as the Lam Tsuen River. Even within a family, e.g., Thiaridae, there are differences in the timing of recruitment. Ultimately, any explanation of snail seasonality must include the effects of temperature as well as spates or scouring, but other factors — such as evolutionary history — may contribute to the patterns observed.

A knowledge of evolutionary history helps us to understand the life cycle of crabs of the genus *Eriocheir* (the hairy-clawed or mitten crabs) which spend most of their lives in freshwater, but must to return to the sea to release their eggs which hatch into planktonic larvae. These crabs belong to the Grapsidae, a primarily marine family. Although the genus *Eriocheir* is the only grapsid colonizer of fresh water, the mode of reproduction is typical of marine crabs and persists in mitten crabs as an evolutionary relict forcing them to undertake seaward breeding migrations. It is during the autumn mating season, when the gonads are ripe, that the Chinese mitten crabs (*Eriocheir sinensis*) are collected for food from the Yangtze Basin. Although they are most delicious when lightly cooked, mitten crabs are intermediate hosts of the lung fluke (*Paragonimus westermani*) and gourmets risk parasitic infection if they consume half-raw crabs. The local mitten crab, *Eriocheir japonicus* (Fig. 6A), is smaller than *E. sinensis* and is not generally collected in Hong Kong for human consumption.

True freshwater crabs never return to the ancestral marine environment. In Hong Kong, they include semi-terrestrial *Nanhaipotamon hongkongense* and the hillstream species *Cryptopotamon anacoluthon* (Potamidae: Fig. 6C and D), as well as *Somanniathelphusa zanklon* (Parathelphusidae: Fig. 6B) which is typical of lowland habitats. All three species produce large, yolky eggs, and the planktonic larval stages typical of marine crabs are suppressed so that hatchlings resemble a tiny adult. The female holds the eggs beneath the abdomen, where they may be incubated for several weeks. Brooding females are reclusive and retreat into burrows or cavities beneath large stones. Juveniles are released near the onset of the wet season in Hong Kong, and production of young during the monsoon seems to be typical of tropical Asian freshwater crabs.

Among the shrimps, *Macrobrachium* species are found in completely fresh water as well as high-salinity brackish water, and these habitat differences have implications for their life cycles. *Macrobrachium hainanense* (Palaemonidae) is common in hillstreams far from the sea and spends its entire life in fresh water. *Macrobrachium nipponense*, by contrast, tolerates brackish water and occurs in tidal shrimp ponds (as in the Mai Po Marshes) as well as in larger rivers and reservoirs. This species can breed in fresh or brackish water, and breeding takes place during the warmer months. Reproduction by stream-dwelling *Macrobrachium hainanense* seems to be stimulated by rising water temperatures and egg-bearing females are found in May and June. Unlike *Macrobrachium nipponense*, however, *Macrobrachium hainanense* shows direct (or abbreviated) larval development lacking the planktonic stages seen in the former species. The same pattern of direct development and breeding seasonality is seen in the small shrimps which abound in most Hong Kong streams: *Caridina lanceifrons* and *Neocaridina serrata* (Atyidae). Suppression of planktonic larval stages and direct development into a miniature adult, similar to the pattern seen in freshwater crabs, may be an adaptation to reduce washout of larvae during spates which can occur during the summer breeding season. An alternative or supplementary explanation is that freshwater

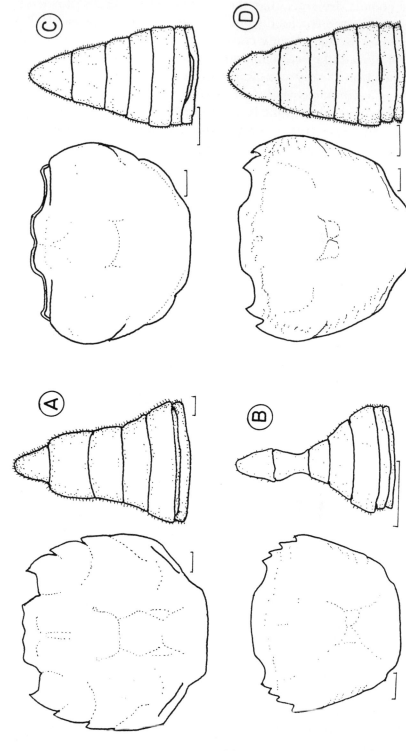

Fig. 6 The carapace and abdomen of male specimens of Hong Kong freshwater crabs: A, *Eriocheir japonicus*; B, *Somanniathelphusa zanklon*. C, *Nanhaipotamon hongkongense*; D, *Cryptopotamon anacoluthon*. Scale line = 4 mm. Drawings by David Dudgeon.

shrimps and crabs produce larger eggs (and hence larger offspring) than their marine relatives because of the paucity of food for planktonic larvae in streams.

Most data on the seasonality of insects in Hong Kong streams concerns dragonflies and damselflies (Odonata). All of the species which have been investigated in detail — the damselfly *Euphaea decorata* (Euphaeidae), and the dragonflies *Zygonyx iris* (Libellulidae), *Heliogomphus scorpio* and *Ophiogomphus sinicus* (Gomphidae) (Fig. 7) — complete one generation per year. Adult flight periods of *Euphaea* begin in late April or early May and adults are abundant

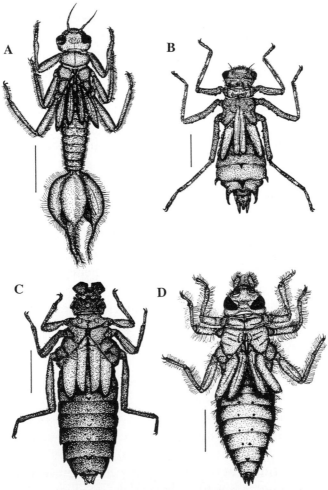

Fig. 7 Hong Kong Odonata (dragonfly and damselfly) larvae: A, *Euphaea decorata*; B, *Zygonyx iris*; C, *Heliogomphus scorpio*; D, *Ophiogomphus sinicus*. Scale line = 5 mm. Drawings by David Dudgeon.

until September, with a few individuals persisting until October. Data on other damselflies, although less detailed, indicate a similar pattern: *Rhinocypha perforata* (Chlorocyphidae) adults occur from May to September, but a few individuals may be seen in April, October and November. *Mnias mneme* and *Neurobasis chinensis* (Calopterygidae) adults emerge earlier in March. The flight season of the latter species continues throughout the summer, but *Mnias* adults are rarely seen after the end of April. Like *Euphaea*, *Zygonyx* dragonfly adults emerge from April to June with the flight period lasting at least until August. Emergence of *Heliogomphus* and *Ophiogomphus* dragonflies peaks in May, but these adults do not remain close to emergence sites, and thus the duration of the flight season is difficult to determine with certainty. However, oviposition by some gomphids takes place in July. At least one other gomphid (*Paragomphus* sp.) emerges in spring and oviposition follows soon after. While most of the new *Paragomphus* generation will not be mature until the following spring, a small proportion of the population of larvae grow quickly enough to emerge in early October. After mating they contribute to a second generation of dragonflies within a single year.

The seasonality of these Hong Kong Odonata is typical of tropical stream-dwelling dragonflies and damselflies. Adult emergence begins before the onset of the summer monsoon, which may reduce damage or wash-out of large larvae during summer spates. Timing of emergence could, however, reflect an influence of rising water temperatures on larval maturation. This pattern does not apply to marshland damselflies, which may produce two or more generations per year with adults emerging in spring give rise to a summer generation which grows rapidly (due to high water temperatures) to emerge as adults that produce a further generation of overwintering larvae which emerge as adults in spring. An additional summer generation may be possible in species which complete larval development rapidly.

While the seasonality of some Hong Kong Odonata is relatively straightforward, this is not true of at least one species — the migratory *Pantala flavescens* (Libellulidae). These insects migrate across the Old World tropics, and huge swarms arrive in Hong Kong during early summer. Some of these individuals mate and oviposit in temporary pools and ponds, but the swarms soon move on. The dragonflies return in early autumn, at the end of the wet season, when swarms are joined by emerging larvae that have developed during the summer months. Some weeks later the throngs depart and are not seen again until the following year. The phenomenon of migration by *Pantala* is well known, but the distance flown by individuals, their flight routes, and the length of adult life are not known.

The seasonality of Hong Kong caddisflies (Trichoptera) varies between species, and this disparity is shown well by a comparison of Stenopsychidae and Macronematinae (Hydropsychidae) (Fig. 8) which, by virtue of their size and (in the latter case) bright colours, are conspicuous components of the caddisfly fauna. Both groups are widely distributed in Asia: the Stenopsychidae, which attains its greatest diversity in north-temperate Asia, decrease in species richness towards the equator, while the Macronematinae are well represented in tropical Asia and the genus *Polymorphanisus* is found nowhere else. Among the Hong Kong Macronematinae, *Polymorphanisus astictus* has the longest flight period, lasting from late April until mid-September. *Macrostemum fastosum* begins flying at the same time as *Polymorphanisus*, but becomes scarce in July, while *Macrostemum lautum* adults are abundant for around two weeks in mid May only. Adult *Stenopsyche angustata*, by contrast, occur throughout the year

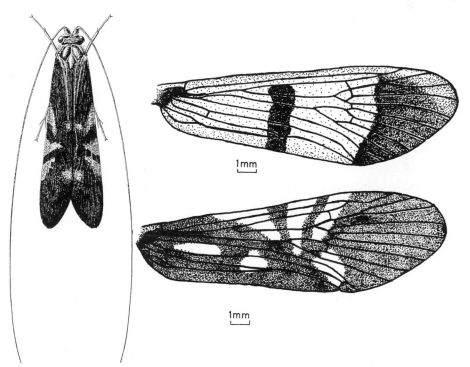

Fig. 8 Caddisfly (Trichoptera) adults: general view of an adult *Macrostemum lautum* (Hydropsychidae); forewing of *Macrostemum fastosum* (upper) and forewing of *Macrostemum lautum* (lower). Drawings by David Dudgeon.

and there is no evidence of seasonality in emergence. Because Macronematine adults are most numerous when daily air temperatures exceed 25°C and water temperatures are above 20°C, it is tempting to suggest that early summer emergence is a response to rising water temperatures, although there is no direct evidence for this. In view of the fact that stenopsychids are primarily a north-temperate group while the Macronematinae (especially the genus *Polymorphanisus*) are more typical of warm running waters, it is possible that macronematine emergence may be confined to the summer months in Hong Kong's seasonal tropical climate, while the 'cool-adapted' Stenopsychidae could fly throughout the year.

Like the variation seen in the seasonality of different caddisfly species, Hong Kong mayflies (Ephemeroptera) include species whose adults fly throughout the year (some Heptageniidae), others which emerge during the summer (*Ephemera spilosa*: Ephemeridae, *Isonychia kiangsinensis*: Oligoneuriidae), and at least one species which shows markedly synchronized emergence over a few days in mid-April (*Ephemera pictipennis*). A lack of clear seasonality or synchrony of larval growth is characteristic of families such as Heptageniidae, Leptophlebiidae and Baetidae, and some of these mayflies are polyvoltine (i.e., completing several generations per year) with periods of extended recruitment and a long period of adult emergence. This pattern is usual in streams where water temperatures, as in Hong Kong, remain above 15°C for a major portion of the year. Because high water temperatures provide the potential for polyvoltinism among Hong Kong mayflies, it is interesting that not all mayflies exhibit such cycles.

A strategy of emergence in late spring or early summer is common to certain Odonata, Trichoptera and Ephemeroptera, and such seasonality involves relatively large species. Significantly, small larvae are less liable than larger ones to be crushed and mutilated by rolling stones during spates. Concordance of life cycles among large species from three insect orders supports the suggestion that spate-induced mortality is an important selective pressure, and this factor may be the ultimate (or evolutionary) cause of timing of emergence in some stream insects. If this is correct, then the proximal (or direct) cause of emergence (imposing a degree of synchrony in the population response) could be a physiological response to rising temperature and/or increasing day length.

One determinant of seasonality which deserves closer consideration is photoperiod, because species with highly synchronized emergence (such as *Macrostemum lautum* and *Ephemera pictipennis*) fly on almost the same calendar days each year. Actual values of temperature and rainfall on any particular calendar day vary widely from year to year, but day length does not. This factor can thus serve as a cue for synchronized emergence on the same date each year. Such synchronization could enhance mating success in low-density populations.

Seasonality of aquatic animals: vertebrates

While we know a considerable amount about breeding seasonality in freshwater invertebrates, there are fewer data concerning the seasonality of fishes, amphibians or aquatic reptiles in Hong Kong freshwaters. For many species, breeding begins in spring. The presence of large numbers of tiny loaches (Balitoridae and Cobitidae) and gobies (Gobiidae) in hillstreams during July and August indicates that reproduction in these fish species is initiated on or before the onset of the summer monsoon. This timing could alleviate the effects of flooding during reproduction by fishes, but mode of development may play a role here. One Hong Kong goby, *Ctenogobius duospilus*, belongs to a genus of fishes (treated as *Rhinogobius* or *Tukugobius* by some specialists) that inhabits a range of river and stream habitats in Asia. Hong Kong *Ctenogobius* produce a small number of large eggs within which the larval swimming stages are passed. The newly-hatched larvae are thus well-developed benthic fish (Fig. 9) with the same body form as the adult. This life-cycle has been derived from the type shown by other freshwater gobies which produce a great number of small eggs that hatch into free-swimming larval stages. Development in these species takes place in lakes or estuaries, although the adult habitat is lowland streams and rivers. In this context it is significant that the benthic loach *Noemacheilus fasciolatus*, as well as other local Balitoridae, also produces a small number of unusually large (approximately 2 mm-diameter) eggs, and the direct development shown by stream gobies parallels that of freshwater crabs and shrimps in Hong Kong.

Among Hong Kong's rich amphibian fauna, newts (*Paramesotriton hongkongensis*: Salamandridae) occur in hillstream pools and are reported to breed in the cooler weather from November to February. They lay about 120 eggs which are attached to submerged objects. Development to hatching takes three weeks or more, and the larvae

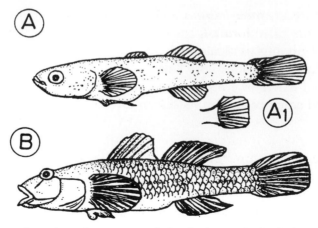

Fig. 9 *Ctenogobius duospilus*: A, a newly-hatched juvenile (body length = 7.5 mm) and, A1, a ventral view of its pelvic fin; B, an adult (body length = 45 mm). Drawings by David Dudgeon.

bear external gills which persist for several months until metamorphosis into the air-breathing adult form. Many of the 22 species of local Anura (frogs and toads) breed during early or late spring, but several species have a lengthy breeding period. Interestingly, the brown wood frog (*Rana latouchii*) which, like the newt genus *Paramesotriton* is at the southern limit of its range in Hong Kong, begins breeding in winter. Tadpoles of a few species of frogs (e.g., the common spiny frog, *Rana exilispinosa*) can be seen in all months, but this could reflect slow larval growth during the cooler months rather than year-round breeding.

Most local anurans become inactive at temperatures below 13°C, but begin calling and mating activity as temperatures and humidity rise in spring. Typically, males of the common Asian toad (*Bufo melanostictus*) are the first to be heard. While much information remains to be gathered, it is probable that reproduction by frogs which breed in standing waters and marshy environments is stimulated primarily by rainfall and rising water levels in spring. For those species which inhabit streams, population size will reflect the interaction of rising temperatures and humidity (which stimulate breeding) and mortality of tadpoles during spates. This conclusion accords with generalizations arising from studies of frogs elsewhere in Asia: rainfall appears to regulate reproductive patterns in areas characterized by a pronounced dry season and, the more northerly the environment, the greater the number of factors which control frog breeding.

Among the reptiles associated with streams are the terrapins *Chinemys reevesi*, *Cuora trifasciata* and *Platysternon megacephalum*. The breeding habits of the former species, which is relatively common and inhabits ponds, marshes and reservoirs as well as streams, are reported to involve pairing in spring followed by oviposition of four to six eggs in June. All of the local terrapins hibernate (or, at least, become inactive) at temperatures below 15°C. Among lizards, the waterside skink (*Tropidophorus sinicus*) is common in and around upland streams and produces three to six live young (i.e. not eggs) in early spring. The water snakes include species of the genera *Natrix* and *Opisthotropis*, which seem to be confined to rocky hillstreams; *Opisthotropis lateralis* (the bicoloured stream snake) is reported to lay eggs in June. *Enhydris chinensis*, a lowland water snake, produces live young (up to 13 in a single brood) during the summer.

To sum up, the diversity of seasonal patterns among freshwater animals in Hong Kong emphasizes the fact that reproductive timing and population dynamics will reflect an amalgam of different influences, including those that arise from a species' evolutionary history, habitat (swift hillstreams versus slow-flowing lowland streams), and vulnerability to spate-induced mortality. Thus, while rising spring temperatures may stimulate reproduction in certain species, it is naïve to expect that this factor alone will be sufficient to account for the range of patterns observed. It is unlikely that we will be able to untangle all of the causative links or be able to make statements that will be true for all species, or even for related species (such as all bivalves, or all fishes). Nevertheless, a preliminary statement about seasonality is that many inhabitants of hillstreams breed in the spring and summer, although whether this is a physiological response to rising temperatures (proximal cause), or an evolutionary adaptation to avoid spate-induced mortality (ultimate cause), or both, is unclear. However, it is striking that some aquatic insects emerge on almost the same calendar date each year, regardless of year-to-year variations in rainfall and temperature, and there is a parallel with the rather rigid timing of leaf replacement in certain terrestrial plants (e.g., *Sapium*). This lack of response in the timing of life-cycle events to fluctuations in weather conditions suggests that seasonality is genetically fixed in some species. Presumably, the individual organism's response (oviposition, leaf replacement, or whatever) is stimulated by an external factor, but the trigger is more likely to be day length variations, which are the same every year, than changes in the weather.

5

Succession and Climax

Succession

Compared with most animals, vascular plants are long-lived, sedentary organisms. It is therefore tempting to view vegetation as merely the static matrix through which animals move. Plants are not immortal, however, and the seed stage is highly mobile. Vegetation does change and often with surprising rapidity. Seasonal changes in vegetation and individual plants have been discussed in Chapter 4. Here we are concerned with longer-term changes in the structure and species composition of vegetation and the associated animals. These changes are termed succession.

It is most convenient to restrict the term succession to unidirectional vegetation change, using other terms for fluctuations and cyclic changes. Ecology textbooks distinguish between primary succession on surfaces, such as newly-exposed rock, which have not previously borne vegetation, and secondary succession in areas where the previous vegetation, but not the soil, has been destroyed. In practice, however, this distinction is far from absolute. Succession on highly-eroded hillsides, for instance, shows some features of both types. Primary succession on totally new substrates is unimportant and unstudied in Hong Kong and will not be considered further.

Early studies of succession emphasized the orderliness and predictability of the changes. The distinguished American ecologist,

Frank Clements, compared primary succession to the life history of an organism and secondary succession to the healing of a wound. Current ecological theory, in contrast, emphasizes the role of individual species biology and of chance. Succession occurs because every local flora consists of plant species with many different combinations of ecological characteristics. When a site is made available for colonization by the destruction of the previous vegetation, different species have different probabilities of arriving at the site because of differences in dispersal ability or in their ability to persist through the disturbance. Among the species that do arrive, there will be variations in establishment success, growth rate, competitive ability, shade tolerance and so on. The result of all these differences between species is the progressive change in structure and species composition that we recognize as succession. In parts of the world where the local flora consists of only a few plant species, succession will proceed in an orderly and predictable manner. In areas such as Hong Kong, however, with a rich flora and many plant species sharing similar ecological characteristics, the particular species present will depend to a considerable extent on chance and on the composition of the surrounding vegetation. In these circumstances, detailed predictions are impossible.

If the physical environment remained constant and there was no further major disturbance, a site would eventually be occupied by all the species in the local flora which could persist indefinitely under such conditions. This final, stable plant community can be termed the climax community for that particular site. In practice, even in the absence of man, long-term changes in climate and rare natural catastrophes, combined with the slowness of plant migration, prevent this final equilibrium from ever being reached. The theoretical climax under a given set of environmental conditions may still, however, be a useful concept for the description and understanding of vegetation change.

Before the arrival of man, Hong Kong was covered in a dense, species-rich rain forest, inhabited by a diverse vertebrate and invertebrate fauna. We can only speculate about the detailed floristic composition of this forest because no large remnants survive, either in Hong Kong or in adjacent parts of South China. The tree families Lauraceae and Fagaceae were probably dominant, judging by their prominence in the regional flora. The importance of the Fagaceae is supported by the prominence of genera of this family (*Castanopsis, Lithocarpus* and *Quercus*) in samples of fruits and seeds from an archaeological site at Penny's Bay, Lantau Island, dated at about 6000 years ago. The samples consist of only exceptionally woody fruits and seeds so the absence of the typically thin-walled seeds of the Lauraceae does not mean that this family was not present near the site. The other species identified were *Choerospondias axillaris* (the hog plum), an unidentified species of *Elaeocarpus*, *Gnetum montanum*, *Pinus massoniana* and *Schima superba*, all of which can still be found in Hong Kong today.

Hong Kong is well within the typhoon belt and this probably had a significant influence on forest structure. The impact of typhoon winds is variable: temporary defoliation is the most widespread type of damage but near the centre of major typhoons large trees may be uprooted or snapped off. The canopy is opened up, allowing invasion by light-demanding tree species which are otherwise confined to large tree-fall gaps. Thus, even before the start of human impact, Hong Kong's forests may have had some features usually associated with disturbance.

Fire and cutting

The original forest has now disappeared almost completely and been replaced by a variety of secondary vegetation types. However, forest remains the potential climax vegetation for Hong Kong's climate and any area protected from disturbance will eventually revert to forest. The succession to forest is halted or reversed by anything that prevents the formation of a closed, woody canopy. In modern Hong Kong, hill fires are the major factor maintaining non-forest vegetation. A single fire eliminates fire-sensitive species and sets back the growth of those which survive. Repeated fires favour those plants which recover fastest. The growth habit of grasses, which grow from the protected base of the leaf, makes them particularly tolerant of repeated burning and accounts for the large extent of grassland on Hong Kong hillsides. Most areas of grassland still retain at least scattered shrubs, with *Rhodomyrtus tomentosa* apparently the most fire tolerant. These shrubs regenerate after fire from the woody stem base but very frequent burning will kill them, probably by exhausting energy reserves and preventing regrowth.

Until a few decades ago, the cutting of hillside biomass for fuel was probably more important than fire in many areas of Hong Kong. The village grasscutters were usually women and a contemporary observer early this century described them as looking, from a distance, like 'miniature haystacks wandering on the mountain-side'. In the short term, cutting seems to have less impact than burning, because it is difficult to cut the plants right down to ground level. However, the nutrients in cut biomass are removed completely from the system so the long-term impact is likely to be more detrimental. Grazing by cattle may also have been (and probably still is) of local importance in inhibiting succession but grazed areas are usually burned as well to encourage fresh young shoots, and thus the direct effects of cattle alone are hard to evaluate.

Cutting and burning have probably affected all accessible areas of hillside for at least the past couple of centuries and probably much longer. The complete absence of native squirrels and the paucity of specialist forest vertebrates in general, suggests that forest destruction was almost complete. Even during periods of maximum human impact, however, remnants of the original forest flora (and the more tolerant elements of the fauna) must have survived in damp ravines, on cliffs, and in other rocky areas that are impervious to fire and not worth the effort of cutting. Woodlands were also preserved or established behind villages for *feng shui* (Chinese geomancy) reasons. These remnants provide the ultimate seed source for development of forest when the hillsides are protected.

Seed dispersal

For a plant species to participate in succession it must first arrive at the site. Most common herbaceous species on Hong Kong hillsides have small, dry seeds or fruits that are apparently dispersed by wind. Only a few herbaceous species (e.g., *Dianella ensifolia*) have fleshy fruits with seeds that are swallowed and dispersed internally by animals. In contrast, more than 85% of the common trees and shrubs have fleshy fruits. Although the tiny, plumed seeds of some herbaceous plants may drift for kilometres, wind-dispersal is inefficient for the larger seeds of most woody species, few of which travel more than 15 m from the parent plant. As a result, wind-dispersed shrubs and trees, such as *Gordonia axillaris*, tend to have highly-clumped distributions where they occur and are completely absent from large areas of apparently suitable habitat. *Pinus massoniana*, which dominated many hillsides before its recent decline (see Chapter 8), might appear to be an exception, but it has been so widely sown and planted for centuries that its original distribution is impossible to determine.

Most of the fleshy-fruited species are dispersed by birds. By far the commonest resident frugivorous birds on Hong Kong hillsides are the Chinese and crested bulbuls (*Pycnonotus sinensis* and *P. jocosus*) and the white-eye (*Zosterops japonicus*). Other common and at least partly-frugivorous residents are the hwamei (*Garrulax canorus*) and the various laughing-thrushes (*Garrulax* spp.), but many other species eat fruit at least occasionally. A number of the migrant birds which arrive in the winter (see Chapter 4: Bird Migration) also eat some fruit, particularly the thrushes and robins but also, surprisingly, some of the flycatchers (*Ficedula mugimaki* and *F. zanthopygia*) and the Chinese bush warbler (*Cettia canturians*).

Birds have no teeth and typically swallow fruits whole. This means that, unless they can peck out small bits, their choice of fruits is limited by their gape width — the largest object the bird can swallow. The gape limit of the bulbuls is 12 to 13 mm, while that of the white-eye is only about 8 mm. Almost 90% of the common fleshy fruit species are less than 13 mm diameter and thus potentially available to bulbuls, while more than half are available to white-eyes. This matching of the requirements of the common fruiting species and the common frugivores does not necessarily imply prolonged coevolution, in which the two groups of species have been major selective influences on each other. A simpler explanation, given the recent creation of these secondary plant communities, is that the bulbuls 'selected' and dispersed suitable fruit from the available regional flora and only these species are now common. This is supported by the existence in Hong Kong of a number of rare species with larger fleshy fruits.

Hong Kong's two species of fruit bat, the cave-dwelling *Rousettus leschenaulti* and the tree-roosting *Cynopterus sphinx*, may be important seed dispersal agents but their role has not yet been studied in any detail. Bats have teeth, and thus complex fruit processing in the mouth is possible, with varying consequences for seed dispersal. Typically, a fruit bat carries a single fruit away from the fruiting tree to a nearby 'feeding roost' — the branch of a tree, the underside of a palm leaf or somewhere similar. The bat processes the fruit while hanging upside down, holding it in one foot. The indigestible parts (the skin and, usually, the seeds) are dropped under the feeding roost and only the fruit pulp swallowed. However, the tiniest seeds cannot be separated from the pulp and are swallowed with it. Most seeds 'dispersed' by fruit bats end up under another tree — rarely an ideal site for establishment and growth. Those small enough to be swallowed, in contrast, are usually defecated in flight and thus widely-dispersed. Elsewhere in the tropics, this habit makes small-seeded, bat-dispersed species particularly important in the early stages of forest succession, but this does not appear to be the case in Hong Kong.

The most important group of plants with known, bat-dispersed species in Hong Kong are the figs (*Ficus* spp.). Fruit bats share some smaller-fruited species, such as *Ficus microcarpa* and *F. superba*, with frugivorous birds but are probably the only seed dispersal agents for others, including two common species (*F. fistulosa* and *F. variegata*) which bear figs on the trunk and branches. Ripe figs of these bat-dispersed species are green (tinged with red in *F. variegata*) and do not attract birds. These two fig species fruit heavily in early summer when the fruit bats are breeding and little other fruit is available. It is therefore not unreasonable to suggest that neither the exclusively-frugivorous bats nor the bat-dependent figs could survive in Hong Kong without the other member of this partnership. Bats will also eat the very similar figs of *F. hispida* but, as most are borne on short, trailing branches near ground level, it is more likely that they are 'targeted' at ground-living mammals, such as pigs, deer or civets. However, we have never found seeds of this species in the faeces of any of these animals.

The small number of common plant species with fruits too large for birds to swallow whole and which are not eaten by bats must have other means of dispersal. Some of them, such as *Rhodomyrtus tomentosa* and *Ficus hirta*, have soft fruits containing many small seeds. Birds peck out small pieces from these fruits. Each piece has several seeds, which are thus dispersed. Bulbuls also peck pieces of flesh from the overripe fruits of *Diospyros morrisiana* but, in this case, the birds seem to avoid the four large seeds so no dispersal is effected. This *Diospyros* and several other big fruits with few, large seeds (such as the gymnosperm climber, *Gnetum montanum*) seem to be dispersed largely by civets (*Paguma larvata* and *Viverricula indica*). Civets climb well and have no difficulty with large fruits because of their wide gape and strong teeth. Most fruits they eat seem to be swallowed after little, if any, chewing and the seeds appear undamaged in the faeces. A single civet defecation may contain 50 or more large seeds. Civets also eat some small-seeded species, such as *Rhodomyrtus* and figs, and are extremely fond of *Elaeocarpus sylvestris*, which is just within the bulbul gape limit. In these cases, birds are probably far more important as dispersal agents because they deposit seeds in small numbers over a wide area while civets leave many seeds, sometimes many hundreds, in a single clump. Civets seem to prefer to defecate in open places, a habit which may have been of benefit to many plant species when Hong Kong was covered in forest. Today, however, it means many seeds end up on rocks, paths, grave sites and other unsuitable sites.

Three fairly common plants in young successional forest do not seem to be dispersed by either birds, bats or civets. Two of these, the tree, *Garcinia oblongifolia*, and a large climber, *Melodinus suaveolens*, produce very large fruits (up to 10 cm diameter in *Melodinus*) with the edible part protected by a thick, resinous outer peel. The third species, *Artocarpus hypargyreus*, has very large but unprotected fruits. Over much of Hong Kong, most fruits of these three species fall to the ground and rot. In parts of the New Territories where macaques (*Macaca* spp.) are present, however, the crops of all three fruits are completely consumed. With *Garcinia* and *Melodinus*, the monkeys remove part of the peel with their incisors while holding the fruit in their hands. They then scoop out the flesh and seeds with their teeth and discard the rest of the peel. The seeds are not swallowed, but spat out individually after separation from the fruit pulp in the mouth. This would not make for very effective seed dispersal were it not that macaques have large cheek pouches in which they store excess food. After stuffing these pouches with pulp and seeds, the animals move on, spitting seeds as they go. The seeds of *Artocarpus* are also spat out.

Seed-spitting seems to be a general characteristic of fruit processing by all species of macaques. Only the smallest seeds (less than 4 mm diameter) are habitually swallowed intact and usually only unripe seeds are chewed up and digested. Macaque faeces from Hong Kong sometimes contain a few larger seeds but these are probably swallowed accidentally. As with the civet scats, macaque faeces containing hundreds of small seeds should not be seen as an example of successful dispersal. Although one or two seeds may sometimes survive and grow, the parent plant would undoubtedly do better, on average, if the same number of seeds was spread between a much larger number of bird (or bat) droppings.

It is hard to believe that macaques are the only seed dispersal agents for these three species, as monkeys have not been recorded in most of the Territory during the last 150 years. Man is an alternative agent for the edible fruits of *Artocarpus* and *Garcinia* — treating them exactly the same way as the macaques do — but not for *Melodinus*, which is reputed to be poisonous. Perhaps civets eat the occasional *Melodinus*?

To complete this account of dispersal, one final type of fruit has to be mentioned, although it is not important in succession. All trees in the family Fagaceae (oaks, chestnuts, etc.) as well as a few genera in other families (*Camellia*, *Styrax*), have large, dry fruits or seeds without an outer fleshy layer. Elsewhere, fruits of this type are dispersed by birds and rodents which store them for later consumption. If the animal stores more than it can consume, forgets some, or dies, then the seeds can germinate. The plant, in effect, pays with some of its seeds for the dispersal of others. This may seem a risky and inefficient means of dispersal but, in North America and Europe, where the Fagaceae are dispersed by specialist birds, members of this family are among the most effectively dispersed of forest trees. The Fagaceae are also important in succession in many tropical highland areas. In Hong Kong, in contrast, the Fagaceae are dispersed poorly, if at all. Large crops of fallen acorns and chestnuts can be found under trees in Tai Po Kau Forest and elsewhere. Macaques consume some species and rodents destroy others, but there is no evidence that any are dispersed. Two species of *Styrax* are the only plants with fruits of this type that occur regularly in secondary vegetation. They have smaller fruits than the Fagaceae and may be dispersed by rodents. The jay (*Garrulus glandarius*), the bird that disperses acorns in Europe, is an occasional winter visitor to Hong Kong, but it is rare and, in any case, many of our local Fagaceae have fruits that are much larger than European acorns. We suspect that the original dispersal agents of these fruits, possibly squirrels or other large rodents, are now extinct. Whether this is true or not, the obvious lack of dispersal of Fagaceae in present-day environments in Hong Kong makes the family a good indicator of older, possibly primary, forest. There are no Fagaceae in most areas of forest known to be secondary, and the exceptions are mostly either planted trees or individuals confined to isolated ravines and rocky areas from which they may never have been eliminated.

Secondary forests

Effective seed dispersal is essential for a plant to participate in forest succession, but getting there is only part of the problem. The plant must also be able to establish itself in competition with the species already present at the site. Plant species diversity increases rapidly in the early stages of succession until grass is finally eliminated and a closed, woody canopy formed. From then on, not only must new invaders establish in deep shade, but all the species already present that are unable to keep up with the rising canopy must be shade tolerant or they will be eliminated.

Hong Kong's flora seems to contain relatively few species that are both well-dispersed and shade-tolerant. As a result, diversity declines after formation of a closed canopy. A few new species are added in the understorey but more are lost as most of the previously established shrubs are shaded out. The young secondary forest is typically dominated by *Schefflera octophylla* and several very similar-looking species of *Machilus*, although other species are locally important. The native pine, *Pinus massoniana*, was the most widespread pioneer species until its recent virtual elimination by the pinewood nematode (*Bursaphelenchus xylophilus*) (see Chapter 8). It is not clear whether the pine acted as a 'nurse', encouraging the establishment of broad-leaved trees, or if it inhibited their growth by shading, but there is no doubt that much of the broad-leaved secondary forest in Hong Kong today, which now covers almost 10% of the Territory's area (Table 1, Fig. 10), started out under pine trees sown or planted after the war.

Schefflera, although not a tall species, persists a long time in the shade of the rising canopy, dying slowly, branch by branch. Eventually, the canopy consists largely of *Machilus*, with a scattering of other species, including *Acronychia pedunculata*, *Diospyros morrisiana*, *Elaeocarpus sylvestris* and *Garcinia oblongifolia*. Under the canopy, in the understorey, the shrubby *Psychotria rubra* is always present, often accompanied by *Ardisia quinquegona* and species of *Symplocos*. The young secondary forest contains few herbaceous plants but there is usually at least one species of *Alpinia* and one or two species of ferns.

Comparisons with old aerial photographs show that this stage can be reached in 40 years on sites with a reasonable depth of soil. Few areas of forest in Hong Kong are older than this because of the drastic cutting for fuel during and immediately after the Japanese Occupation. Thus, the nature of the later successional changes, as with the original forest cover, is open to debate. Most of the present canopy dominants are not regenerating, despite abundant fruit production. This is probably because their seedlings require more light than is available in the understorey. In fact there are few seedlings of any potential canopy species in the understorey of most secondary forest areas in Hong Kong.

If left undisturbed, the present canopy dominants will eventually reach maximum size and start to die. If more shade-tolerant species have become established in the understorey by this time, they are

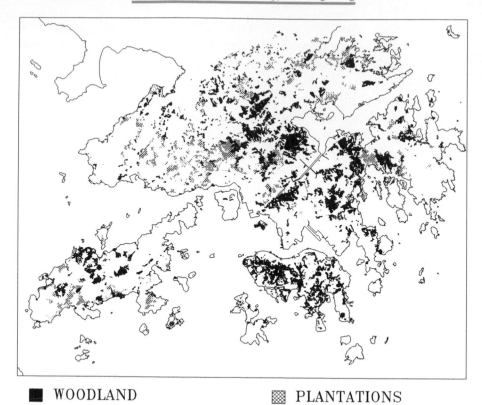

■ WOODLAND   ▦ PLANTATIONS

Fig. 10 A map of Hong Kong showing the extent of secondary forests (woodland) and plantations.

likely to replace the light-demanding pioneers. If not, the present dominants will presumably replace themselves by regeneration in tree-fall gaps and the forest will continue to be dominated by pioneers indefinitely.

It is important to note that succession is not simply a biological process and, particularly on highly eroded sites, the rate of soil development may often be a limiting factor. Tall, closed forest cannot develop on the shallowest soils while any significant increase in soil depth will take hundreds or thousands of years.

Feng shui woods

A possible future direction of succession is suggested by some of the *feng shui* woods behind old villages. These woods are preserved or planted for reasons of *feng shui* (literally, 'wind-water'), the Chinese system of geomancy. *Feng shui* woods in the Territory range from those that are obviously dominated by planted trees, often of ornamental or economic species, through all possible intermediates to a minority which appear more or less natural. Unfortunately, this appearance may be deceptive as human intervention, by either cutting or planting, which ended a few decades ago would be very difficult to detect at the present day. Using these woods as a guide to future successional trends thus risks circular argument. Moreover, because of their locations, many *feng shui* woods are probably on better sites than the typical post-war secondary forest. The early successional stages on such sites may have been very different from those observed on hillsides today.

The best *feng shui* wood in the Territory, reputed to be 400 years old, is in Shing Mun Country Park at the north end of Jubilee Reservoir. Here the dominant trees are *Endospermum chinensis*, *Pygeum topengii*, *Sterculia lanceolata*, *Sarcospermum laurinum*, *Cryptocarya chinensis* and *Machilus* spp. *Cryptocarya chinensis* is regenerating abundantly in the understorey. This area invites comparison with Ding Hu Shan Biosphere Reserve, 86 km west of Guangzhou, where with a similar climate and flora, the oldest forest (again believed to be more than 400 years old) has a canopy dominated by *Castanopsis chinensis*, *Schima superba*, *Cryptocarya concinna* and *C. chinensis*, with the *Cryptocarya* species dominating the regeneration. *Pygeum topengii* and *Sarcospermum* are also common in this forest but *Endospermum* is apparently absent.

There are a number of other *feng shui* woods in Hong Kong of similar species composition and probably similar age to the Shing Mun site. Among the best are those behind Sheung Wo Hang village in the north-east New Territories and along Nam Fung Road (around the Hong Kong Electric cable tunnel portal) on Hong Kong Island. This latter site is shown as wooded on nineteenth-century maps and is probably the only old woodland on the island.

Although undoubtedly old, the composition of these *feng shui* woods is not what one would predict for the original forest cover of the region. Despite their diversity, they are dominated by species that can be dispersed by the existing animal fauna, while poorly-dispersed groups, such as the Fagaceae, are rare or absent. This suggests that they are secondary and raises the interesting question: are there any primary forest remnants in Hong Kong today?

Montane woodland

The most likely candidates for remnants of the original forest cover of Hong Kong are all at high altitude (above 500 m). On the upper slopes of Tai Mo Shan, Ma On Shan, Sunset Peak and Lantau Peak, there are patches of woodland which may be primary. These woodlands are in damp valleys, above cliffs or among boulders — all situations likely to exclude fire. But would they be safe from cutting? These areas are 'remote' today (meaning far from roads) but the human population distribution was more even in the past. Abandoned tea terraces provide proof of cultivation at high altitudes and there are clear signs of old terraces within many forest patches. These woodlands must also have been tempting sources of firewood for tea-growers, charcoal-burners and upland villagers.

The strongest support for the primary (but not necessarily pristine) nature of some of these woodlands comes from their floristic composition. Although the prominence of the Fagaceae (*Castanopsis*, *Lithocarpus*, and *Quercus*) and species of *Camellia* may be partly a reflection of the cooler climate at these altitudes, these are all trees with hard, dry and, probably, poorly-dispersed fruits. They are not found in known secondary woodlands. Indeed, it is hard to imagine any natural process by which the huge (three or more grams) acorns of *Lithocarpus corneus* could be dispersed across the wide expanses of grassland between valleys on Tai Mo Shan. The presence of this and similar species must surely be a relict of once continuous forest cover. This does not mean, however, that considerable disturbance in the past is ruled out. There are surprisingly few large trees, and most individuals are multi-trunked, suggesting recovery from past cutting.

The montane woodlands are also interesting for biogeographical reasons. Many species which are ubiquitous in woodlands at lower altitudes become rare or absent at high altitude. The highest patches have no *Acronychia pedunculata*, *Aporusa dioica*, *Garcinia oblongifolia* or *Psychotria rubra*. These are all genera which are characteristic of equatorial lowlands. Conversely, other species which are rare or absent at lower altitudes, such as *Ligustrum lucidum* and several species of *Camellia*, are common in the montane forests.

Plantations

As well as natural woodlands, Hong Kong has a large expanse of
artificial plantations, covering almost 5% of the area of the Territory
(Table 1, Fig. 10), almost all established within the last 45 years.
Although the Chinese pine (*Pinus massoniana*) was the most widely
planted species for centuries, the extensive pre-war pine plantations
were lost during the Japanese occupation and post-war plantings of
this species have largely succumbed to the introduced pinewood
nematode (see Chapter 8). Most surviving plantations consist of a
single exotic (i.e., non-native) species. The most widely planted species
is Brisbane box (*Lophostemon confertus*, formerly known as *Tristania
conferta*) from eastern Australia, but there are also large areas of slash
pine (*Pinus elliottii*) from the southeast United States, acacia (*Acacia
confusa*) from Taiwan, and swamp mahogany (*Eucalyptus robusta*)
and paper-bark (*Melaleuca quinquenervia*) from eastern Australia.
Several other exotic species have also been planted in smaller areas.
The most widely-planted native species, again in single-species stands,
is *Schima superba*.

Plantations differ from natural forests of similar age in several
important ways. Most obviously they have a much lower plant diversity.
Not only is the canopy formed of a single tree species but the
understorey is usually poorly-developed, as a result of the dense shade
and, possibly, slowly-decomposing leaf litter. Plantations are also much
simpler structurally than natural secondary forests. These two features
— lower species diversity and simpler structure — tend to reduce the
diversity of the animal fauna of plantations. No quantitative
comparisons have been made in Hong Kong, but the reduction in bird
diversity on moving from natural secondary forest to young plantation
at Tai Po Kau is immediately obvious. None of the common plantation
species produces fleshy fruit and they are chosen for planting, in part,
for their resistance to insect attack. The simpler structure and general
absence of a dense understorey and climber tangles must also reduce
the availability of nesting sites for birds. Old plantations, in contrast,
typically have a well-developed native understorey and seem rich in
wildlife.

Production of timber is not a commercially-viable proposition in
Hong Kong, but the reafforestation programme is justified on grounds
of erosion control and landscaping value. On the most eroded sites,
exotics may be the only trees that can be established but their use in

areas which can support mixed native woodland is much harder to justify. It may be possible to use exotics as 'nurse' species to promote succession by providing protection to invading natives. Pines seem particularly effective in this role. However, typical planting densities in Hong Kong are too high and the resulting dense canopy is more likely to inhibit natural succession. The quickest way to establish a native woodland is to plant mixed, native species but, on good sites, natural succession after fire control may be almost as fast.

Animals in succession

It is simplest to consider the development of the animal community separately from that of the plants because animal populations typically adjust much more rapidly than plants to changes in the environment. The two processes are far from independent, however. Not only is the composition of the animal community to a large extent determined by the structure and species composition of the plant community, but the animal species present have a major impact on plant succession, most obviously through seed dispersal but also, presumably, through herbivory, pollination and so on. Unfortunately, little is known currently about the factors influencing the composition of terrestrial invertebrate communities in Hong Kong, and thus most of the following discussion is concerned only with vertebrates.

The vertebrate fauna of Hong Kong grasslands contains relatively few species. Grazing mammals, which contribute to much of the vertebrate diversity and biomass of natural tropical grasslands and savannas, such as those of East Africa, are entirely absent from South China. This no doubt reflects the recent origin of this habitat in a region that has apparently been forested throughout the Pleistocene. The vertebrate fauna of grasslands in Hong Kong is dominated by seed and insect-eating birds and rodents. Seed-dispersing frugivorous birds are absent as there is no food available for them. Flocks of bulbuls may cross grassland areas when flying between patches of more suitable habitat but, as birds normally defecate only while perching, seeds will not be deposited often. This must slow the initial rate of succession when grassland areas are first protected from fire. Civets, however, do enter grassland to hunt rats, and any faeces deposited there may contain seeds brought from woodland areas several hundred metres away.

Increasing shrub cover attracts more vertebrates by providing both cover and food. The new arrivals include frugivorous birds with guts full of seeds. Bulbuls seem to reach peak density in the dense shrubland stage as (probably) do other fruit-eating birds. The same may be true of mammals but we have no comparative data. As the shrub canopy closes the open, grassland habitat is eliminated while a new, protected, interior habitat is created. Neither of the common bulbul species nor white-eyes penetrate much below the canopy but, as succession proceeds, a new bird fauna arrives to exploit this habitat, including laughing-thrushes (*Garrulax* spp.), the violet whistling thrush (*Myiophoneus caeruleus*) and coucals (*Centropus sinensis* and *C. bengalensis*). In winter, these species are joined by several species of true thrushes (*Turdus* spp.) and robins (*Luscinia* spp.), as well as a number of Palaearctic warblers (*Phylloscopus* spp.).

Finally, some bird species seem to be confined largely to sizable areas of semi-natural woodland, such as Tai Po Kau Forest. Here one can see birds such as the great barbet (*Megalaima virens*), which requires mature trees for nesting holes, the scarlet and grey-throated minivets (*Pericrocotus flammeus* and *P. solaris*), and the emerald dove (*Chalcophaps indica*), which are rarely observed elsewhere. Much of the original forest interior vertebrate fauna is, however, undoubtedly extinct.

6

Land and Water

The stream and its valley

Inhabitants of streams and rivers must contend with the unidirectional flow of water from the upper reaches down to the sea. This has the consequence that events upstream affect those downstream, so that running-water habitats do not consist of a series of distinct autonomous sections but rather an interconnected continuum. A second distinctive feature of streams is that they are embedded in terrestrial landscapes and cannot be considered in isolation from their surroundings. Water entering a stream or river has flowed over the land surface or percolated through the soil so that its dissolved and suspended loads reflect, to some degree, the nature of the landscape. The edge to area (or volume) ratio of lotic (running water) habitats is a good indicator of terrestrial influences on the aquatic environment. The combined extent of both banks along a stream course presents a large expanse over which lateral movement of material into a relatively small aquatic habitat can take place. These edge- or bank-effects must be taken into account when conservation and management strategies for streams are formulated. While a sufficiently large woodland patch may be more or less independent of surrounding habitats (except near the margins), the extent of the banks of streams and rivers and the fact that they drain the surrounding land ensures a close link between streams and their valleys. Maintaining the integrity of land-water interactions is therefore a necessary element of stream conservation and management.

An additional link between the stream and its valley arises because the water in the channel is not confined within the banks but is continuous with the ground water underlying the surrounding land; this overlap of land and water is most extensive in lowland flood-plain rivers with alluvial valleys. Thus, streams are related to their surroundings in three spatial dimensions: interactions are longitudinal along the stream tributary network to the estuary, vertical with the groundwater, and lateral with the banks, soil water and riparian vegetation. There is an important fourth dimension to consider in understanding stream communities — that of time, with seasonal (e.g., monsoons, floods) and longer-term (e.g., climatic transformation, acidification) changes having short-term behavioural or longer-term evolutionary effects.

Despite the four-dimensional nature of stream ecosystems, the unidirectional flow of current is the major factor underlying the distribution and abundance of lotic organisms. The stream can be thought of as a one-way channel along which water, as well as suspended and dissolved materials, both organic and inorganic, travel. The persistence of stream communities depends on the ability of the constituent organisms to collect and retain transported materials which serve as food for animals, as a substratum for microbes, or as nutrients for plant growth. At the risk of oversimplifying, many attributes of the stream community will reflect the balance between the biological potential for collection and retention of organic matter and nutrients, and the downstream export of such material by the current.

The influence of physical factors

Stream organisms are influenced by a range of interacting abiotic factors, the most important of which are current velocity, flow pattern and discharge, substratum characteristics (including particle size and unevenness), temperature, and dissolved oxygen. The nature of the physical environment in streams reflects the effects of current velocity on the movement of inorganic particles, and the composition of the stream bed in a given section (e.g., the proportions of sand, gravel and boulders) affects the type of organisms which are able to live there. Because the physical characteristics of a stream change along a continuous gradient from headwaters to mouth, so too will the stream community alter and thereby exhibit a pattern of longitudinal zonation. In addition, however, there will be variation or 'patchiness' in the physical habitat at any single point along the stream course. For example, where the stream bed is composed mainly of boulders, smaller rocks and gravel will fill up the interstices of the substratum between them. Further complexity is added because sand and fine particles accumulate in isolated pockets sheltered from the current along the stream margins and under, or in the lee of, large rocks and boulders. There is a pattern to some of this substrate variation because physical characteristics change markedly and in a predictable fashion across the stream bed. Current velocities are highest in midstream (often over 50 cm/second) and lowest close to the banks (generally less than 10 cm/second) where there is considerable frictional resistance to water flow. Because the ability of the current to pick up and transport particles decreases with velocity, sediments are coarse with a predominance of large particles in midstream, and there are increasing proportions of fine particles in patches of substratum close to the banks.

Inter-patch differences in sediment characteristics are the main contributors to environmental heterogeneity within the stream, and affect the distribution of benthic invertebrates which crawl or sprawl on, and burrow in and among, the particles making up the stream bed. Many benthic animals therefore show characteristic distribution patterns across the stream, and some species are most numerous close to the banks (e.g., the leptophlebiid mayfly, *Thraulus bishopi*) while others (including another leptophlebiid, *Isca purpurea*) are more abundant in midstream. These patterns may change seasonally because spates associated with summer monsoonal rains scour the stream bed, disrupt across-stream sediment gradients, and wash fine particles downstream. In effect, inter-patch heterogeneity is reduced with the result that bank-side and midstream sediments become more similar. Across-stream gradients in particle size and substrate characteristics are reestablished as discharge volumes decline and stabilize during autumn and winter, and the distribution of benthic animals returns to pre-spate patterns as sediment heterogeneity is restored.

Many benthic animals do not experience the full force of the water flow because boundary-layer effects (a result of the resistance to water flow caused by the friction of the stream bed) are prominent, and current velocities decrease markedly with proximity to the bottom sediments. In addition, dead-water spaces occur downstream of obstructions to flow. These rather static films and pockets of water serve as microhabitats for many benthic organisms, and provide a refuge from the full force of the current. As a result, distribution patterns are determined less by the direct effects of current and more by the effects of water flow on the transport and deposition of inorganic particles. Hence the current influences stream organisms indirectly by way of an effect on substratum. The boundary layer does, of course, become thinner with faster rates of flow, and it is not possible for benthic animals to escape entirely from the effects of current. Indeed, hydraulic variables must play an important role in the metabolism, feeding, and behaviour of the stream fauna, especially those filter-feeders that depend upon the current to bring them food. We still have much to learn about the complex effects of water flow on their ecology.

Longitudinal zonation

Man has long known of the association of particular animals and plants with particular river sections, and the use of streams as a source of food must have spurred the first attempts at the identification of such zones. The earliest schemes, based on the presence of dominant fish species, originated in Germany towards the end of the nineteenth century and were later extended or modified to apply to other parts of Europe. Fine-tuning German river-zonation schemes to match circumstances in different countries highlighted one significant drawback in the use of fishes (or any faunal group) to identify different river zones — the key species might be absent from a given region due to geographic or historical reasons. In addition, transitions in faunal assemblages are more often gradual than abrupt, so that discrete zones cannot be identified readily and longitudinal changes in different elements of the stream community do not always occur at the same points along rivers. This highlights the fact that stream inhabitants are distributed in response to a complex of physical factors. If individual species respond uniquely to combinations of these factors, their patterns of longitudinal zonation will differ. For this reason it is unrealistic to expect synchronized changes in community composition at well-defined boundaries along the river continuum.

To what extent do these remarks apply to the Hong Kong running waters? None of the larger streams and rivers in Hong Kong are pristine as they are all influenced variously by pollution, extraction of water, and channelization. It is unwise, therefore, to consider longitudinal change and zonation without taking notice of human influences. The picture is complicated further by seasonal fluctuations in pollution loads which result from the diluting effect of increased stream discharge between April and September when over 80% of the annual rainfall occurs. Patterns of longitudinal zonation which do arise are due, in part, to Hong Kong landforms which ensure that most streams have steep stony courses with rather short sections of gently-flowing water in the lower valleys. Exceptions (such as the Sheung Yue River, Shek Sheung River and Ng Tung River which all drain into the Sham Chun River) pass through densely-populated rural areas and close to burgeoning New Towns. They are polluted by agricultural, industrial and domestic effluents.

The water chemistry of unpolluted streams reflects Hong Kong's igneous geology and the impoverished character of the thin acidic soils which cover most hillsides (see Chapter 2). Stream waters are soft, slightly acidic and nutrient poor, although dissolved oxygen levels and silicates are generally high. Downstream dilution of dissolved substances in pristine rivers results from a decrease in the input area (the river bed and banks) in relation to the increasing size and volume of the main stream. In Hong Kong, however, downstream increases in nutrients and reductions in dissolved oxygen are a typical result of organic pollution. There is a general trend also towards increases in the amounts of suspended organic matter along the course of local rivers. This can be attributed to drifting filamentous and unicellular algae sloughed from the stream bed, fragments of aquatic plants, and inputs of agricultural wastes. In contrast, the bulk of suspended material in the headwaters is detritus fragments derived from terrestrial leaf litter.

Among local rivers, the Lam Tsuen River offers us the best insight into the zonation pattern of Hong Kong's stream fauna. The river drains a broadly circular catchment of approximately 18.5 km² situated west and inland of Tai Po. The highest point in the watershed is at 800 m elevation, and the area consists mainly of volcanic rocks with an alluvial valley floor. The headwaters arise to the southwest of Tai Po on the slopes of Tai Mo Shan and are marked by a series of spectacular waterfalls. The river then flows through hillside shrubland and forested ravines, and then into cultivated land devoted to market gardening as well as pig and fowl rearing. (A slight improvement in water quality recently may be attributable to the introduction of government restrictions on pig farming.) Approximately 2.5 km from the original river mouth (now displaced seawards by land reclamation), the course changes from a southwest-northeast to a northeast-southwest direction. Slightly downstream of this, the flow is blocked by an inflatable fabridam from where water is pumped, through tunnels, into Plover Cove Reservoir. Only a small fraction of the river discharge avoids this diversion and, in 1984, the section below the dam was channelized, destroying the natural habitat. The account presented here is based on information which predates that destruction, at which time the river was moderately polluted along part of its course. Accordingly, the picture that emerges from this example is some way from the ideal of a pristine habitat.

Despite despoliation by man, the Lam Tsuen River supports a diverse fauna, including animals of various habits from a range of phyla. Longitudinal zonation is exhibited by most denizens, even neustic species such as heteropteran bugs and gyrinid (whirligig) beetles which are associated with the water surface. The Heteroptera include representatives which walk upon on the surface film, of which the Gerridae (water skaters) are most conspicuous, and others such as belostomatids, water scorpions (Nepidae) and backswimmers (Notonectidae) which hang suspended from the water surface. Some of these animals (e.g., the large water skater *Ptilomera tigrina*) are characteristic of turbulent headwater sections, while the backswimmer (*Enithares*) is found in pools between such reaches. Belostomatid bugs and certain water skaters (*Limnogonus fossarum* and *Gerris paludum insularis*) are typical of the quieter downstream sections where they may be associated with aquatic macrophytes (i.e. non-microscopic plants). Most heteropteran bugs are carnivores which feed on terrestrial insects that become trapped on the surface film; their habit is such that they straddle the boundary between air and water.

Longitudinal zonation is apparent also among the nekton, i.e., those actively swimming animals that can move about in the water against the force of the current. In streams, this niche is occupied largely by fishes, although beetles (some Dytiscidae and Hydrophilidae) swim freely in the water column in slow-flowing weedy reaches. There has not yet been a complete fish survey published for any Hong Kong river, and our knowledge of the ecology of local species is hampered because of this. Nevertheless, it is clear that certain species are typical of headwater sites in the Lam Tsuen River. The minnow-like fishes of stony upstream reaches inhabit pools, and include the cyprinid *Parazacco spilurus*. In fast-flowing, turbulent reaches, some fishes have abandoned the nektonic habit for a secondarily bottom-dwelling (benthic) way of life that has involved modification of the body, pelvic and (sometimes) pectoral fins. Among them are the herbivorous balitorid loaches *Pseudogastromyzon myersi* and *Liniparhomaloptera disparis*, as well as predatory gobies (*Ctenogobius duospilus*) and loaches (*Noemacheilus fasciolatus* and *Oreonectes platycephalus*). These species are intolerant of organic pollution, and are replaced by different fishes, including exotic species, further downstream.

In low-gradient streams, and the lower course of relatively unpolluted rivers, fishes with a wide variety of body forms and habits are represented. They include the eel-like loach *Misgurnus anguillicaudatus*, the common carp (*Cyprinus carpio*), catfishes such as *Parasilurus* spp. (Siluridae) and *Clarius batrachus* (Clariidae), the half-banded barb (*Capoeta semifasciolata*) and the minnow, *Zacco platypus* (Cyprinidae). Other minnows, such as *Yaoshanicus arcus*, are confined to the lower reaches of a few small streams in the east of the Territory, while *Acrossocheilus wenchowensis beijianensis* (Fig. 11) (a strikingly-banded small carp) is found only in Tung Chung Stream on Lantau Island. Other fishes found in the lower course of local rivers are species which involve both a marine and freshwater stage in their life cycles. They include *Plecoglossus altivelis* (a small, salmon-like fish which is now extremely rare), *Eleotris oxycephala* (Eleotridae) and *Ctenogobius giurinus* (Gobiidae). These species are in decline because a combination of pollution and flow-regulation projects have obstructed the routes between fresh and salt water. (The conservation status of local freshwater fishes is discussed in more detail in Chapter 9.) In slow-flowing streams, the paddy-field eel *Monopterus albus* (Synbranchidae) occurs in sites supporting dense growths of aquatic macrophytes. It has an auxiliary intestinal respiratory organ and takes

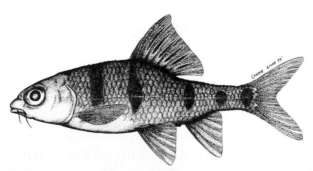

Fig. 11 *Acrossocheilus wenchowensis beijiangensis* (Cyprinidae) is known from nowhere else in Hong Kong except Tung Chung Stream on Lantau Island; body length up to 110 mm. Drawing by Chong Dee-hwa.

atmospheric air from the water surface. The paddy-field eel shares this ability with the paradise fishes (*Macropodus opercularis* and *M. concolor*: Belontiidae), which have an accessory breathing apparatus in the head. Despite its respiratory adaptations, *Macropodus opercularis* is not confined to living in oxygen-poor water and occurs in a variety of aquatic habitats.

In addition to native fishes, the lower course of the Lam Tsuen River is host to an array of alien or exotic fishes. Among them, the African mouth-brooding cichlid *Oreochromis mossambicus* (one of a number of species known as 'Tilapia'), has spread across tropical latitudes as a result of its use in aquaculture. Four species of live-bearing Central American poeciliid fishes have also established self-sustaining populations: *Poecilia reticulata* (the guppy) and *Gambusia affinis* (the mosquito fish) were introduced to Hong Kong in an attempt to control mosquito larvae, while *Xiphophorus helleri* (the swordtail) and *X. variatus* (the variable platy) are escapees from the aquarium fish trade. The importance of these and other exotic or alien species will be considered further in Chapter 8.

The bottom-dwelling or benthic animals constitute the richest faunal assemblage in the Lam Tsuen River, and, in addition to the fish mentioned above, include turbellarians (predatory flatworms), annelid worms and leeches, gastropod and bivalve molluscs, decapod crustaceans (crabs and shrimps), and a predominance of insects. Most of the latter are larval forms which mature into winged, terrestrial adults. The insects with this amphibiotic life cycle are mayflies (Ephemeroptera), damselflies (Odonata: Zygoptera), dragonflies (Odonata: Anisoptera), stoneflies (Plecoptera), caddisflies (Trichoptera), beetles (Coleoptera) and true flies (Diptera), as well as a few moths (Lepidoptera: Pyralidae), riffle bugs (Heteroptera: Naucoridae) and fish-flies (Megaloptera). Over 120 different taxa have been recorded from the bottom sediments alone but, because of the difficulty of separating the larvae of closely-related species, the actual total must be greater than this. Many of the common species have yet to be described or named by experts.

Benthos species richness declines along the river from headwaters to mouth. However, organic pollution in the lower course increases food availability for some tolerant species, allowing them to build up dense populations. The downstream changes are due also to transformations of the benthic environment, particularly differences in substratum characteristics (especially particle size) along the river. It is perhaps not surprising that upstream sites with a wide range of particle sizes and a physically complex or heterogenous substratum support a greater range of benthic species. Further downstream, fine-grained sediments fill the interstices between larger particles causing a loss of microhabitat which could have been occupied by burrowing animals. The addition of fine organic particles (derived from agricultural wastes) to the river may restrict water movement within the bottom sediments, so reducing the availability of oxygen. A more direct effect may be to clog the gills and feeding apparatus of some animals, thereby contributing further to the downstream decline in benthos diversity.

A degree of compensation for the downstream loss of benthic species arises from the presence of trailing terrestrial vegetation (grasses and the like) as well as floating or attached aquatic macrophytes whose roots, stems and leaves provide important habitats for a wide range of animals. Among these plants, *Hydrilla verticillata* (Hydrocharitaceae), *Ludwigia adscendens* (Onagraceae), and water hyacinth (*Eichhornia crassipes*: Pontederiaceae), grew in profusion in lower Lam Tsuen before it was channelized in 1984. The submerged roots of water hyacinth, an exotic floating plant from South America, provided an important habitat for over 90 species of gastropods, dragonfly and damselfly larvae, heteropteran bugs, beetles and so on; many of these animals did not occur on or among the bottom sediments. The extent of growth of aquatic plants in Hong Kong rivers is, to a great extent, determined by the flow volume and, at times of high discharge, the plants may be washed out to sea. During the dry season, by contrast, plants grow in profusion and may almost choke the water course.

We do not intend to give an exhaustive account of the longitudinal distribution of individual members of the benthic fauna of the Lam Tsuen River, though it is worth noting that some groups of animals, the Plecoptera (stoneflies), are confined largely to one section of the river (the headwaters), while other groups show a succession of species replacements along the course. The mayflies (order Ephemeroptera) epitomize the latter group, and can be used to illustrate zonation patterns and the factors underlying them. Prior to channelization of the lower course, 31 mayfly species had been recorded from the Lam Tsuen River; 21 of these species were abundant. Mayflies did not occur on or among the bottom sediments of the river below the fabridam, although some (mainly *Cloeon*) did live among the roots of floating plants. All but one of these species were restricted to certain parts of the river course, and even the exceptional species was numerous at downstream sites only.

Respiratory adaptations can provide an explanation for variations in the longitudinal zonation of different mayfly species, and this is especially applicable to the Baetidae (Fig. 12A) which constitute the most species-rich mayfly family in Lam Tsuen River (Fig. 13). Genera which inhabit turbulent headwater sites (*Baetiella* and *Indobaetis*) have small, immobile abdominal gills which cannot beat to create a respiratory current and rely on a strong flow of water to carry dissolved oxygen over the body surface. Under conditions of respiratory stress, *Baetis* spp. (which occupy a range of habitats along the Lam Tsuen

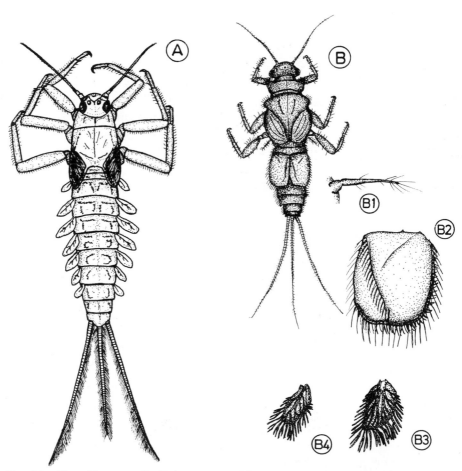

Fig. 12 Hong Kong mayfly (Ephemeroptera) larvae: A, *Baetis* sp. (Baetidae) showing the lateral abdominal gill lamellae — body length (excluding 'tails') = 8 mm; B, *Caenis* sp. (Caenidae) — body length (excluding 'tails') = 6 mm; B1, the first abdominal gill; B2, the modified second abdominal gill or operculum; B3, the third abdominal gill; B4, the sixth abdominal gill. Drawings by David Dudgeon.

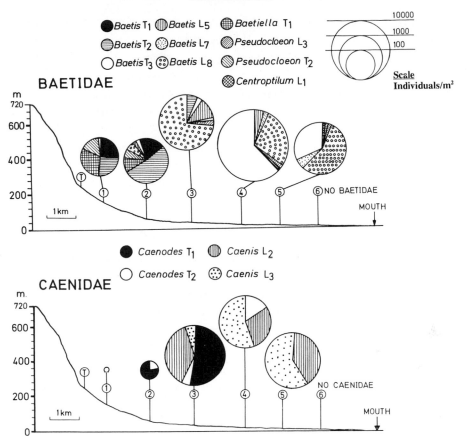

Fig. 13 Longitudinal zonation and abundance of baetid and caenid mayflies during the wet season in the Lam Tsuen River. Letters and numbers are used to designate different species within the same genus as the larvae of many Hong Kong mayflies have yet to be described named by scientists.

River) beat their gills to create a water current and facilitate respiration, while genera typical of weedy areas in the lower course, such as *Cloeon*, have moveable gills with expanded, lobed lamellae (that may be paired) which present a large surface for gaseous exchange.

As well as species-specific zonation patterns within families such as the Baetidae, clear differences in the distribution of mayfly families along the river are apparent. One or more members of the family Baetidae are numerous at all sites where mayflies can be found, but representatives of the Caenidae are scarce upstream yet abundant further downstream (Fig. 13). Many caenids inhabit silty substrata which are typical of the lower course of rivers; their distribution reflects the morphology of their abdominal gills and the way that they beat (Fig. 12B). In baetids, the moveable gills draw a current over the body, bringing water in from the front and sometimes the side also. Caenids have a reduced first gill; the second forms an robust operculum overlying the abdomen and succeeding gills which are thin and delicate. When they beat, these moveable gills cause a current from side-to-side across the body. If the animal rests on the substratum with one side slightly raised, it can draw water in from that side without disturbing the silt and so avoids gill clogging.

Respiratory adaptations of the type described above attain special significance in rivers experiencing organic pollution. While these habitats remain well-oxygenated during the day because of the photosynthetic activity of benthic algae, oxygen levels decline at night when photosynthesis ceases while respiration continues. Oxygen is consumed also by the microbes associated with organic wastes, so that night-time sags in oxygen level can create critical conditions for animals living on the stream bed.

The zonation of different families and species of mayflies in Lam Tsuen River (Fig. 14) reflects the physical and chemical changes that take place along the course of the river, and the modification of these environmental characteristics by seasonal changes in pollution load. In essence, high flows during the wet season flush out the river so that pollution is reduced and much of the course is available for colonization. Increased pollution loads during the dry season eliminate many species from their preferred habitats downstream and, for example, truncate mayfly distribution along the river. Recolonization of the lower course by species of Baetidae and Caenidae (among others) takes place during the summer monsoon when river discharge increases (Fig. 14). Overall, the longitudinal zonation of Lam Tsuen River benthos is closest to that of a pristine water course during the wet season. Elevated organic loads associated with decreasing river discharge during the dry season eradicate some species from the lower course, but the greater food availability and gentler flows in the river at this time can favour the establishment of dense populations of tolerant species which may include introduced or alien taxa (see Chapter 8). Thus the diverse natural community of the river penetrates relatively far downstream during the wet season but, during the dry season, much of the lower course is dominated by a relatively species-poor assemblage of pollution-tolerant organisms.

Where it is possible to divide the river course into sections on the basis of faunal complement, each zone that is identified will include organisms with similar responses to physical factors, but, although they live in the same place, these organisms may not interact or be

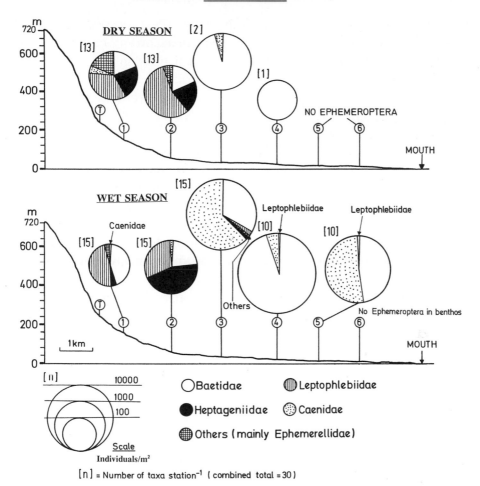

Fig. 14 Longitudinal zonation and abundance of mayfly (Ephemeroptera) families in the Lam Tsuen River: wet season versus dry season.

linked functionally. Without a functional basis, however, schemes of river zonation and classification can have little explanatory power or predictive value. For example, it is clear that some benthic invertebrate families display a downstream replacement of species along the Lam Tsuen River, while others are confined mainly to the headwaters or lower course. Without a knowledge of their functional roles, it is not clear whether the substitution of one species by another has any effect on community functioning. For example, subtle changes in community composition and representation of individual species could result in major changes in community functioning and energy flow to consumers high up the food chain or, alternatively, the replacement of one species by another with a similar ecology may have negligible effects on such processes. One way to distinguish between these possibilities is to employ a functional classification of animals according to similarities in resource utilization.

A functional approach

Running-water benthos the world over is dominated by insect larvae, in terms of both numbers and species richness. The diversity of benthic communities, coupled with the difficulties of identifying larvae to the species level, has encouraged stream ecologists to develop a functional characterization of these animals according to how they feed. While the scheme was developed for insects, it is applicable, in principle, to other aquatic animals such as crabs, shrimps, snails and fishes. Functional classification of invertebrates has the advantage of reducing the difficulty of dealing with insufficiently-known families, since closely-related species often fall within the same functional group. Simplification of community structural data is another benefit as it facilitates recognition of patterns in ecosystems.

There are four widely-recognized functional feeding groups. The grazer-scraper category comprises herbivores that feed on periphyton (i.e., algae and associated material attached to stones and other submerged objects), including those herbivores which pierce plant tissues or cells and suck out fluids. Shredders are detritivores feeding on large particles, especially decomposing leaf litter derived from the riparian zone; these animals may depend upon microbes on the litter surface for part of their nutrition. Collectors ingest small (< 1 mm diameter) particles and associated microbes, and can be subdivided according to whether the food particles collected are suspended in the water (as in the case of filtering-collectors or filter-feeders) or have been deposited on the substratum (in the case of collector-gatherers). The predators include species which swallow or engulf prey and those which pierce their victims with beak-like mouthparts and inject digestive enzymes before sucking body fluids.

One difficulty with this functional classification is that there may be overlaps between functional groups. Some species, such as certain filter-feeding caddisflies (Hydropsychidae), sieve particles of detritus and algae from the current with the aid of a silken capture net but behave as predators when an animal is caught on the mesh. In addition, some animals may change category as they grow. Despite these complications, many biologists believe that the representation of the four functional components listed above is necessary for community stability in pristine streams and rivers. In running waters receiving organic enrichment, an additional category — deposit-feeders — can include species (such as annelid worms) ingesting fine bottom sediments and the organic material that they contain. Alternatively, these animals may be considered as specialized collector-gatherers.

We began this section by drawing attention to the four-dimensional nature of stream ecosystems. We have seen that the communities of lotic animals change along the longitudinal dimension of the ecosystem, but they are also influenced profoundly by exchanges of material which occur across the lateral dimension. Streams and rivers are flow-through systems, receiving all material passing through them from the surrounding landscape and storing it, exporting it downstream or mineralizing some organic inputs. Members of the functional feeding groups can be viewed as interrelated temporary storage bins for organic compounds which are eventually converted to carbon dioxide. In other words, these organisms consume reduced carbon compounds (i.e., plant parts such as litter) derived primarily from the surrounding land, supplemented by (mostly) algal material produced by in-stream photosynthesis, and bring about their transformation and temporary storage as animal tissues or conversion, via respiration, into carbon dioxide. Linkages between the feeding groups are characterized by changes (usually reduction) in the size of particles. Specifically, shredders consume leaves, bark, and so on and, by the action of chewing and defaecating, comminute the food to produce fine particles that can be gathered or filtered by collectors. The faeces that collectors produce are generally of similar size to the food particles that they ingest and, in turn, their faeces may be ingested by other collectors. Herbivores graze the plant stocks (generally algae) in the stream and, they too, produce faeces which can be taken up by collectors. Microbial colonization of faecal pellets enhances their food value to collectors and may encourage animals to feed on them. Because of this colonization, a proportion of the initial organic input is converted to carbon dioxide by microbial respiration.

The continued cycle of colonization and recolonization of detritus particles by microbes, and their repeated ingestion and defaecation by collectors, can be seen as improving the efficiency with which inputs from the land are utilized. Ultimately, the predators consume shredders, collectors and herbivores, so feeding indirectly upon the detritus and algae which has been assimilated by their prey. The efficiency with which food is transferred between groups, together with the physical characteristics of a particular river reach, will determine the magnitude of downstream loss of organic matter or the 'leakiness' of a stream section.

The River Continuum Concept

How can an appreciation of functional feeding groups and their interactions help us understand river zonation and how stream ecosystems in Hong Kong actually work? The importance of the longitudinal dimension in lotic systems is linked to our understanding of the lateral axis and especially the transfer of materials. These two dimensions have been combined in the River Continuum Concept (RCC) which was developed in North America and links riparian vegetation, aquatic productivity and stream community structure (Fig. 15). The RCC takes account of the gradients of physical variables within a river from headwaters to mouth, envisaging it as a continuously intergrading series of biological adjustments to these gradients. Processes in downstream reaches are linked to those upstream, and predictable longitudinal variations in community organization occur in response to varying hydrological and physical conditions, as well as changes in the resource base. One view is that benthic communities are structured in such a way that they make efficient use of energy inputs. This feature does not arise from some grand co-operative venture on the part of stream inhabitants, but from the tendency of individual species to exploit their environment by maximizing energy consumption.

The RCC has allowed formulation of useful generalizations concerning the magnitude and variation in the organic matter supply, and the effects that this has on the structure of aquatic communities in North American rivers. As a first step, lotic communities are grouped according to channel size into headwaters, medium-sized streams and large rivers. Many headwater streams are influenced strongly by riparian vegetation, which reduces aquatic primary production by shading and contributes large amounts of leaf litter. Shredders are predicted to be co-dominant with collectors in such streams, reflecting the importance of the riparian zone and the detritus derived from it. As stream width increases and shading decreases, the reduced importance of litter inputs coincides with a greater significance of aquatic primary production and the import of fine organic particles from upstream. In medium-

The River Continuum Concept

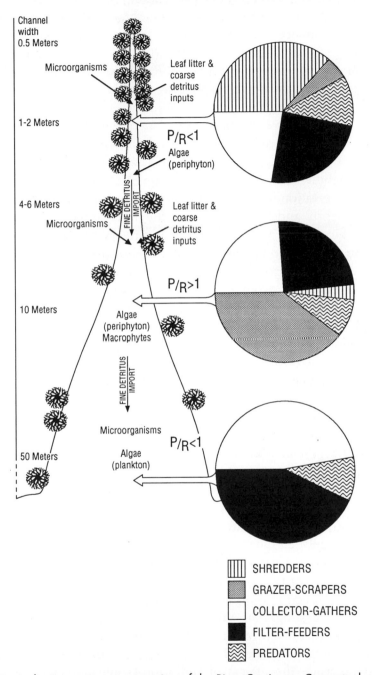

Fig. 15 A diagrammatic representation of the River Continuum Concept, showing the predicted relative abundance of invertebrate functional feeding groups along an unpolluted river. (Further explanation of the Concept and its predictions are given in the text.)

sized streams grazer-scraper biomass is maximized, but collectors are numerous also. The transition from headwaters, dependent on terrestrial inputs, to medium-sized rivers, relying on algal or aquatic macrophyte production, is associated with a change in the ratio of gross primary productivity (P) to community respiration (R). The position at which the stream shifts from heterotrophic ($P/R < 1$) to autotrophic ($P/R > 1$) is dependent upon the degree of shading and thus the form and extent of the riparian vegetation.

Large rivers receive fine particulate organic matter from upstream, and there is a general reduction in detrital particle size as stream width increases. As a result, collector-gatherers and filter-feeders dominate macroinvertebrate assemblages downstream. Although the shading effect of riparian vegetation is insignificant, primary production may be limited by water depth and turbidity; here the river once again becomes heterotrophic ($P/R < 1$). Few Hong Kong rivers are large enough for primary production to be limited by depth, but their lower courses are nevertheless characterized by a heterotrophic community metabolism. This arises from the proliferation of pollution-tolerant species which can exploit the profligate supplies of organic material derived from human activities in the drainage basin.

Although some of the tenets and the universality of the RCC have yet to be evaluated fully, it does highlight the critical linkage between the stream and its terrestrial setting, and places considerable importance upon biotic interactions within the stream community. Data from the Lam Tsuen River are in broad agreement with some predictions of the RCC, but shredders are never sufficiently numerous in the headwaters to co-dominate the benthic community with collectors (Table 2). Further downstream, organic enrichment of the lower course leads to a simplification of community structure during the dry season leading to a dominance of the benthos by deposit-feeders. The summer monsoon flushes out the river and spates cause catastrophic drift which (through wash-out) largely eliminate the deposit-feeders. Consequently, filter-feeders and collector-gatherers become the dominant functional groups throughout most of the river. By contrast, there is no marked seasonal change in the functional organization of the upper-course community (Table 2), suggesting that the alterations which take place further downstream are a result of the interaction between rainfall seasonality and organic pollution rather than a consequence of season alone.

Leaving aside the effects of pollution, some agreement between longitudinal trends in the functional organization of Lam Tsuen River communities and those predicted by the RCC can be seen. There is a

Table 2 Relative abundance (percent) of benthic invertebrate functional groups along the Lam Tsuen River: wet season (W) versus dry season (D). Note that sampling sites are arranged along the river from Site 1 in the headwaters to Site 6 in the lower course.

	Site 1 W	Site 1 D	Site 2 W	Site 2 D	Site 3 W	Site 3 D	Site 4 W	Site 4 D	Site 5 W	Site 5 D	Site 6 W	Site 6 D
Shredders	6	14	4	4	1	0	0	0	0	0	0	0
Grazer-scrapers	14	18	21	14	8	3	8	2	2	0	0	0
Collector-gatherers	36	39	46	41	49	48	48	10	51	6	44	7
Filter-feeders	36	26	24	30	37	41	40	6	34	4	40	5
Predators	8	13	5	11	5	8	4	2	5	2	6	2
Deposit-feeders	0	0	0	0	0	0	0	80	8	88	10	86

tendency towards gradual change in aquatic productivity and functional group representation along the river, but these are affected by organic enrichment due to human activities. The human influence is all-pervasive in Hong Kong, and would seem to present an insurmountable barrier to our use of the RCC to understand local rivers and streams. However, it is possible to ascribe some of the deviations from the RCC which do arise, for example the scarcity of shredders, to localized effects such as the low stature of riparian vegetation in much of the upper Lam Tsuen Valley (although this characteristic, too, is a result of human influence in the past). The riparian influences are worthy of study because they enhance our understanding of streams as components of higher landscape units. If the aquatic community responds predictably to changes in the resource base, then variations in riparian vegetation will modify in-stream processes. In other words, streams with similar physical characteristics and water chemistries will differ in their ecologies according to the vegetation of their valleys. How then does riparian vegetation affect the ecology of local hillstreams?

Terrestrial vegetation and Hong Kong streams

Changes in riparian vegetation alter the relative proportions of energy available from leaf litter and in-stream photosynthesis. Therefore, we would expect a change in community structure across a gradient of shading intensity by riparian vegetation. This is demonstrated by a comparison of four, unpolluted, Hong Kong streams which differ with respect to shading, but otherwise have similar physicochemical characteristics. Tai Po Kau Forest Stream is an entirely shaded site where the canopies of riparian trees interlock over the stream. A second site, Bride's Pool, is also surrounded by trees but the canopies do not shade the entire stream bed. A tributary of the upper Lam Tsuen valley, where the riparian vegetation consists of shrubs and tall grasses with few trees, exemplifies a site almost completely exposed to the sunlight; while a fourth site, Pui O Stream on Lantau Island, is unshaded due to the low stature of the riparian grassland, and dense growths of filamentous algae cover parts of the stream bed.

Differences in the extent of shading by riparian vegetation affects the amounts of detritus in the streams, which are significantly higher in Bride's Pool and Tai Po Kau Forest Stream than elsewhere (Table 3). Periphyton biomass is greatest at the unshaded sites, attaining a maximum in Pui O Stream; both Tai Po Kau Forest Stream and Bride's Pool support sparse algal growths. A clear distinction can be made between shaded streams, with a detrital food base, and streams with an open canopy which contain less detritus and greater amounts of algae. Community composition of benthic animals varies among streams, and only 27% of the total of 126 species are found at all four sites. Tai Po Kau Forest Stream supports the lowest faunal densities but the greatest species diversity (Table 3). There are, however, broad

Table 3 Population densities (number/m²) of benthic invertebrate functional groups in four Hong Kong streams that differ with respect to shading by riparian vegetation. Tai Po Kau Forest Stream was entirely shaded; Bride's Pool was partly shaded; Lam Tsuen River (upper course) experienced limited shading only; and Pui O Stream was unshaded. The effects of shading on the food base of these streams is apparent from the amounts of detritus and algae present at each site. (Note that the units of biomass for detritus and algae are grams and milligrams respectively.)

	Tai Po Kau Forest Stream	Bride's Pool	Upper Lam Tsuen River	Pui O Stream
Shredders	199	49	69	5
Grazer-scrapers	388	593	1131	1046
Collector-gatherers	742	3127	2579	2028
Filter-feeders	833	1044	2154	871
Predators	246	694	483	202
Total population	2408	5507	6416	4152
Total species	94	80	88	70
Detrital biomass (g/m²)	66	39	11	17
Algal biomass (mg/m²)	6	4	18	64

similarities in species complement between Tai Po Kau Forest Stream and Bride's Pool, and also between Lam Tsuen River and Pui O Stream, despite considerable differences between the two pairs of sites. In other words, the species composition of benthic communities in the shaded streams is different from that of the unshaded streams.

Although collector-gatherers and filter-feeders dominate the communities of all four streams, population densities of all functional groups (except predators) differ significantly between sites (Table 3). Grazer-scraper densities are highest at the unshaded sites (upper Lam Tsuen River and Pui O Stream), while shredders are most numerous in Tai Po Kau Forest Stream, where detritus is plentiful. Among grazers, species which pierce and suck the fluids from algal cells (hydroptilid caddisflies) are especially abundant in Pui O Stream where they feed on dense growths of filamentous algae.

Here we have clear evidence for the role of algae, detritus and riparian shading in determining benthic community structure in Hong Kong streams. While human modification of streams and their valleys has made it impossible to use the RCC as a predictive model for changes along a single Hong Kong river, it is nevertheless possible to understand differences among unpolluted hillstreams in the light of the concept.

Land-water interactions: streams

Research on streams in Hong Kong and elsewhere indicates that running waters cannot be considered in isolation from their drainage basins, and this notion is a cornerstone of the RCC. In unpolluted Hong Kong streams, the influence of the terrestrial landscape is a reflection of geology (contributing nutrients) and riparian vegetation (determining the balance between detritus and algae as energy sources). In polluted sections of Lam Tsuen River, transfers of material from land to water are augmented by inputs of domestic and agricultural refuse. These wastes degrade water quality and eliminate certain taxa, while the abundance of a few tolerant species is increased.

We can contrast land-water interactions in the Lam Tsuen valley with those recorded for a near-pristine section of Tai Po Kau Forest Stream. The stream drains a managed nature reserve on the southwestern shore of Tolo Harbour. Although the secondary forest and planted vegetation includes introduced species (see Chapter 5:

Plantations), the majority of plants are native to South China and to this extent the area probably approximates conditions prevalent on many Hong Kong hillsides prior to forest clearance. The reserve is drained by a 3.6 km-long stream which rises 400 m above sea-level and has a forest catchment of approximately 2.5 km². This area is composed predominately of volcanic rocks and the acid soils have a pH of around 5. As we have seen already, the biological environment of Tai Po Kau Forest Stream is influenced profoundly by riparian trees which cast shade and reduce the intensity of light reaching the water surface. Wet-season spates cause substantial reductions (over 90%) in the already sparse algal growths. As a result, the availability of algal food sources for animals in Tai Po Kau Forest Stream fluctuates markedly with time. By contrast, detritus (fallen leaves, bark, twigs, and branches) forms conspicuous accumulations in areas of slow flow, constituting dense packs in places where water is funnelled between large boulders, and standing stocks are always over 100 times greater than algal biomass. Seasonal fluctuations in the amounts of detritus in the stream are rather slight, and detritus carried downstream during spates is replaced by material washed from the forest floor into the stream. Direct leaf fall (from the overhanging closed canopy) into Tai Po Kau Forest Stream exceeds 0.75 kg/m²/yr, in addition to over 0.37 kg/m of stream bank carried in by lateral transport each year.

Although there is some variation between species, leaves are broken down rapidly in Tai Po Kau Forest Stream, with complete disappearance occurring in less than 13 weeks. This is compared to over one year for breakdown of some types of leaf litter in temperate streams. Rapid leaf-litter processing and the large amounts of detritus suggest that algal primary production plays a minor role in the nutrition of benthic animals in Tai Po Kau Forest Stream (see Chapter 7: Detritivores). It is sufficient to note here that one measure of the relative importance of algae and terrestrial inputs as food sources in a stream can be obtained by measuring the energy fixed by in-stream photosynthesis per unit time, and the energy respired by the stream community over the same time. This circumvents many of the problems associated with inadequate characterization and/or quantification of potential food supply, because it is clear that where photosynthesis is less than community respiration the shortfall in energy demand must be made up by inputs from the land. Measurements of community respiration (R) and primary production (P) in a shaded reach of Tai Po Kau Forest Stream have yielded a P/R ratio of 0.17, confirming the dependence upon terrestrial energy sources. In an unshaded stream pool containing abundant algae, a *P/R* ratio of 1.02 was recorded; even at well-lit sites, community respiration almost exceeded primary production.

Streams in forested catchments receive energy subsidies in the form of detritus from the terrestrial environment. In such environments, however, there is a reciprocal movement of material from water to land. Such transfers can occur when the trailing roots of riparian vegetation take up nutrients from solution, and extensive areas of submerged root mats line the bed and banks of many streams. The emergence and nuptial flights of adult aquatic insects may serve also as a pathway for water to land transfers. Such behaviour passes energy and nutrients to the land, because most adult insects enter terrestrial food chains when they are eaten by spiders, birds and bats. Consumption of fish and amphibians (which subsist, at least in part, on stream invertebrates) by water snakes (*Opisthotropis* spp., *Natrix* spp., *Enhydris chinensis*) and birds (including kingfishers, herons and egrets), as well as certain bats (*Myotis ricketti*), could also constitute a mechanism of water to land nutrient transfer.

Complementary to the water to land transfer of material, is the view inherent in the RCC that the communities of unpolluted streams are structured in a way that utilizes the food supply efficiently so minimizing downstream losses to the estuary. The slowing of downstream export in streams is thought to result from capture, ingestion and reingestion of organic particles by filter-feeders, collector-gatherers and other detritivores (as described above). Ultimately, detritus and the nutrients it contains is cycled through the bodies of numerous invertebrates, becoming more completely mineralized with each step in the transfer, thus downstream export is retarded. The process of re-use of detritus in streams has been termed 'spiralling'. The larger detritus can play an additional role in this process, because organic debris dams which obstruct water flow tend to promote the retention of organic particles in the blocked sections.

The idea that downstream losses of nutrients could be significantly retarded (or even reversed) by the directed, upstream flights of adult stream insects is more controversial. The view is that, after emergence, adult female insects fly upstream before laying their eggs, so moving the nutrients their eggs contain uphill in the opposite direction to the flow of stream current. Data from Tai Po Kau Forest Stream and elsewhere provide supporting evidence of this behaviour in some species, but not others. Despite uncertainty over the importance of upstream flights, it is clear that streams can alter the quantity and quality of nutrients and organic material passing through them by uptake and storage of the former, and oxidation of the latter to carbon dioxide. Together, mechanisms that retard the downstream export of nutrients and organic matter, and processes which transfer material from water to land, will have the combined effect that unpolluted streams will tend to be less 'leaky' (losing a smaller proportion of the detritus input to downstream) and will have more symmetrical land-water interactions (involving a significant water to land transfer of material) when compared to organically-polluted or man-modified streams. Tai Po Kau Forest Stream, with two-way land-water interactions, and Lam Tsuen River, with predominately land to water transfers of material, epitomize the opposite ends of this continuum. In essence, the normal functioning of the Lam Tsuen River ecosystem has been disrupted; where such disruptions are severe, they can lead to the collapse of food chains and the loss of species at higher trophic levels — such as fish and other vertebrates — which are typically those of interest to man.

Land-water interactions: Plover Cove Reservoir

Apart from streams, the ecology of another type of freshwater habitat in Hong Kong is influenced profoundly by land-water interactions. Reservoirs, which are developed typically by impounding streams to form a lake behind the dam, experience seasonal fluctuations in water levels. Waters rise during the wet season, and fall during the drier months when inputs from the catchment diminish. On steep reservoir shores, falling water levels expose the underwater portion of the habitat slowly, due to the gradient, thus few aquatic organisms are stranded and die. The freshwater mussel *Limnoperna fortunei* (Mytilidae) is an exception however, as the animal is attached to rocks by byssal threads (secreted by a gland in the foot) and thus cannot follow retreating water levels. Aside from the elimination of *Limnoperna*, water-level fluctuations seem to have few lasting effects on the steep rocky shores which characterize the marginal zone of many Hong Kong reservoirs. In Plover Cove Reservoir, which was formed by damming a marine inlet and transforming it into a freshwater impoundment, falling water levels during the dry season strand molluscs and macrophytes (*Vallisneria spiralis*) on gently-shelving sand-mud shores, although some snails do keep pace with the retreating water line. The stranded molluscs have a limited capacity to withstand desiccation, and only one snail

(*Sinotaia quadrata*; Fig. 5A) can tolerate more than one month out of water. Consequently, recolonization of the newly-flooded marginal zone during the wet season must involve migration of animals from deep water.

On gently-shelving shores, the stranded molluscs and macrophytes represent a transfer of energy and nutrients from water to land. The aerial exposure of marginal-zone muds and the availability of nutrients from the decomposed bodies of aquatic organisms allows the establishment of a terrestrial plants (grasses, sedges and herbaceous Polygonaceae), the extent and diversity of which depends upon the duration of low water levels, but which can reach standing stocks of 60 g/m^2 in ten weeks. Where cattle are present, they graze the vegetation and deposit dung along the shore. When water levels rise (usually in May) inundating the marginal zone, they are accompanied by changes in water chemistry. These changes are confined to the vicinity of gently shelving sites, and do not occur around steep shores. Specifically, nitrogen, phosphorus and organic matter (particulate and dissolved) in the water column increase markedly immediately after the shore has been inundated, but decline as reservoir volume increases during refilling. In situations where light penetration (and hence photosynthesis) is not limited by turbidity, elevated nutrient levels stimulate rapid growth by *Vallisneria spiralis* soon after flooding of the marginal zone.

Inundated terrestrial plants on the shore of Plover Cove Reservoir are consumed rapidly by reservoir fish (especially mud carp such as *Cirrhinus molitorella*) which swim into the inundated area as water levels rise. Within three weeks, these fish can eat 90% of the terrestrial plant biomass. Combined with the release of nutrients from flooded dung and mud, this represents a transfer of material from land to the aquatic habitat, reversing the direction of transfer that took place when invertebrates were stranded by falling water levels. Although aquatic invertebrates in Plover Cove do not feed upon terrestrial plants directly, they may consume carbohydrate-rich fish faeces, and movements of the fish before defaecation would disperse the terrestrial input within the reservoir. Significantly, detritus standing stocks on the mud surface in the marginal zone increase by almost ten times after water levels rise, and food quality of the detritus (in terms of percent organic matter) more than doubles. This transfer of material from the terrestrial environment must enhance food supply to benthic inhabitants of the reservoir.

Water-level fluctuations influence the ecology of Plover Cove, and the effects are magnified by the fact that most benthic animals live in the shallower waters of the reservoir. The influence is comparatively slight on steep rocky shores, but is significant in gently-shelving areas where mass stranding of aquatic organisms and two-way land-water interactions can occur. These reciprocal interactions are functionally similar to the transfers of material associated with Tai Po Kau Forest Stream and, although the reservoir habitat is man-made, it illustrates the pattern of water-level fluctuations seen in the natural lakes of the seasonal tropics. By contrast, one-way (land to water) transfers of material characterize the Lam Tsuen River. Such lopsided interactions seem characteristic of systems perturbed by man, whereas reciprocal interactions between land and water may typify aquatic habitats in a more natural landscape setting.

7

Foods and Feeding

Primary production

Living organisms require energy for their activities and matter for their construction. Only green plants can make direct use of solar radiation as a source of energy and simple inorganic molecules as a source of matter. All other organisms, with the minor exception of some autotrophic bacteria, depend on green plants, the primary producers, for food. The key process in primary production is photosynthesis, which takes place only in the chloroplasts. In most algae, all cells have chloroplasts and can photosynthesize. In higher plants, in contrast, a varying proportion of the plant body consists of non-photosynthesizing living cells — functioning in support, storage and below-ground absorption — and dead cells, serving mainly for support. Largely as a result of this, the photosynthetic output per kilogram of plant biomass (the productivity to biomass ratio) decreases rapidly from aquatic communities, where most plant cells photosynthesize, through terrestrial grasslands and shrublands, with an increasing proportion of non-photosynthetic cells, to forests, where a large proportion of the biomass is dead bark and heartwood. When compared on the basis of area, however, primary productivity is higher in forest and closed shrubland than in grassland because the depth of the canopy means there are more photosynthetic cells per unit area. The productivity of grasslands on Hong Kong hillsides is also limited because of the death of much of the foliage in winter (most trees and shrubs are evergreen) and the often considerable proportion of bare ground.

There have been no measurements of the primary productivity of terrestrial communities in Hong Kong. Net primary productivity (i.e., photosynthetic production minus plant respiration, or the plant biomass that would be available for consumption by herbivores) is of particular interest. Comparisons with other parts of tropical and subtropical Asia suggest that the net primary productivity for mature forest in Hong Kong would be around 2 kg/m^2/yr. This includes only above-ground production; below-ground production, which may be as much or more, is very difficult to measure accurately. The productivity of most existing plant communities in the Territory will be considerably less than this figure because few communities approach the maximum potential leaf cover (leaf area index) of 7–8 m^2 of leaf area per m^2 of ground surface area that is typical of climax forest. Shrubland productivity is probably of the order of 1 kg/m^2/yr — about half that of forest. Measurements of above-ground biomass for local vegetation are scarce also but data on similar vegetation types in Guangdong suggest a range from around 3 kg/m^2 for shrubland to at least 20 kg/m^2 for secondary forest. Shrubland production to biomass ratios are unlikely to exceed 0.3, while in secondary forest the figure would be less than 0.1.

Primary productivity in hill streams is very low. This is partly because of shortage of nutrients in the water. The phosphorus and nitrogen needed by algae (and, indeed, all organisms) to build proteins are present only in minute amounts unless the water has been polluted by sewage, agricultural fertilizers or livestock wastes. One nutrient present in excess is silica, which is used by diatoms (a type of algae) to build their cell walls or frustules. On its own, however, silica does not increase the amount of plant food available to stream animals. Production is limited further by shading. Streams provide a water source for hillside plants throughout the dry season and, even on an otherwise grass-covered hillside, riparian shrubs and trees may shade the stream bed completely. Algae are thus limited by light as well as by nutrients, and stream food chains depend largely upon plant litter derived from trees and shrubs lining the banks (see Chapter 6: Land-water interactions: streams).

Herbivores

The greenness of Hong Kong hillsides suggests a superabundance of food for plant-eating animals. That this is not necessarily so, however, can be seen by imagining oneself having to rely on hillside plants for food. The vast majority of the available plant biomass is too tough, too poisonous or too low in nutritional value. Better-adapted mouth parts and a modified digestive system could overcome some of these problems but the basic fact remains: plant bodies are much poorer nutritionally than animal bodies.

There are two major differences between the cells of plants and those of animals which influence their food value. First, and probably most significant, is the fact that plant cells have thick cell walls. In higher plants these walls are constructed largely of cellulose fibres in a matrix of pectins, hemicelluloses, proteins, lignin and sometimes silica (e.g., 8% of the dry weight of leaves of the tree *Sterculia lanceolata* is silica). These substances impart physical strength and resistance to chemical breakdown; properties which, in turn, make plant tissues difficult to digest. No animal can digest lignin. Cellulose-digesting enzymes are produced by many microorganisms, particularly fungi, but by rather few invertebrates and no vertebrates. Cellulose digestion by animals is usually dependent on the presence of mutualistic microorganisms in the gut. However, even for organisms that can partially digest them, cell walls are a low-quality food with a high carbon to nitrogen ratio which limits their value as a raw material for the synthesis of animal proteins.

In contrast to the cell walls, the contents of plant cells (the cytoplasm) are neither tough nor resistant to digestion. Nevertheless, the ratio of edible cytoplasm to indigestible cell walls in many plants is low, and plants lack the highly-nutritious tissues, such as muscles, which make animal prey such valuable dietary items. Plants make life still more difficult for herbivores by the production of defensive chemicals which deter feeding, inhibit digestion or are directly toxic to animals. The most common types of defensive chemicals are phenolic substances (such as tannins), nitrogen-containing compounds (such as alkaloids, cyanogenic glycosides and non-protein amino acids) and terpenoids, but the total range of substances a herbivore may encounter is immense. The picture is complicated by the fact that some of the chemicals are produced only in response to damage caused by herbivores. Chemical defense is not unknown in animals, particularly soft-bodied, slow-moving ones (such as some caterpillars), but animals generally rely on hiding, disguise or escape for defense — strategies that are unavailable to green plants.

Thus, the bulk of plant biomass is available only to organisms that can either digest cell-wall components or penetrate the cell wall and overcome defensive chemicals, or do both. Animals that can digest cell walls, but are unable to deal with chemical toxins, consume plant material only after it is dead and the toxins have been leached out or decomposed by microorganisms. Although certain beetle larvae, such as the longicorn (or long-horned) and jewel beetles (Cerambycidae and Buprestidae, respectively), and some fungi consume the dead woody tissues inside living trees, most cell-wall specialists are detritivores feeding on dead plant material after it has entered the litter layer. Cytoplasm specialists, in contrast, would be expected to prefer soft, living, plant tissues such as young leaves and shoots, where the mechanical problems of penetrating the cell wall are least. The type of defensive toxin deployed differs among plant species, and thus cytoplasm specialists, unlike detritivores, are confined typically to feeding on one or a few closely-related plant species. This reduces the range of enzymes the animal must produce for detoxification.

Herbivorous animals that can utilize both cell walls and cytoplasm are rare in most communities. This may be because the two types of defense require very different adaptations, but a simpler explanation is that the ability to digest cellulose provides little additional benefit to a herbivore. Most herbivores are probably limited by the availability of nitrogen rather than the carbon or energy that cellulose can provide. The best examples of cellulose-digesting grazers are ruminant mammals, such as cattle, which use microbial fermentation both for detoxification and cellulose digestion. In parts of the world with large areas of natural grassland and savanna, such as East Africa, vast herds of ruminants consume a high proportion of the available biomass. Hong Kong, however, has only secondary grasslands and the only native ruminant is the barking deer (*Muntiacus reevesi*). Little is known about its diet in Hong Kong but it is believed to browse on young leafy shoots as well as consuming fallen fruit when available. Feral or tended groups of cattle (*Bos bovis*) are important grazers in some localities. In many Asian rain forests, leaf monkeys (langurs), with a functionally similar but independently-evolved digestive system, are the arboreal equivalent of ruminants, although they consume a much smaller proportion of the available biomass. Hong Kong has no leaf monkeys today but they may well have been here in the past. The native rhesus macaque (*Macaca mulatta*) does eat considerable quantities of leaves but, with a digestive system little different from our own, cannot digest much cell wall material and has to be selective to avoid poisoning by plant defensive compounds.

The main terrestrial herbivores in Hong Kong are insects, although snails (such as the arboreal forest snail, *Cryptosoma imperator*) may be important in some localities. Caterpillars, which often feed on only one or a few species of plant, dominate the grazer feeding category, but various beetles (such as the leaf beetles, flea beetles and tortoise beetles in the family Chrysomelidae), stick insects (Phasmatodea), and grasshoppers are important. Among the latter, short-horned grasshoppers (Acrididae) are most abundant in open country, along paths and in forest clearings, while long-horned grasshoppers or katydids (Tettigonidae) are more typical of trees and forest understorey plants.

All of the insects mentioned above have chewing, mandibulate mouthparts. Bugs, which are members of the order Hemiptera (especially the suborder Homoptera: cicadas, aphids, spittle bugs and so on), also feed on terrestrial plants but do not chew up the leaves. Instead, they penetrate the plant tissue with their piercing mouthparts and suck sap from the vascular system. Most sap-suckers penetrate the phloem and feed on the carbohydrate-rich phloem sap but some (such as certain spittle bugs: Cercopidae) penetrate the xylem and consume xylem sap, which is much more dilute. The damage done to plants by sap-suckers is hard to quantify, but the diversity and abundance of Hemiptera in Hong Kong suggests that they may consume as much plant biomass as the grazers. Some sap-suckers cause additional damage by secreting toxins in their saliva or by transmitting plant viruses.

Estimates of the percentage of leaf area removed by grazers are usually low. The figure most often cited in the ecological literature for communities elsewhere is less than 10%, and measurements made locally in mixed secondary forest suggest a value as low as 3%. These figures may be underestimates because of limitations in methodology, but it is clear that most of the primary production is not eaten by grazers. Low food quality is one explanation for this, but an alternative view is that herbivore numbers are limited by predation to below a level where plants are damaged severely. These possibilities (which are not mutually exclusive) are still fiercely debated by ecologists.

Much of the primary production on Hong Kong's hillsides is concealed below ground in the form of roots. As with leaves, the finest rootlets are vulnerable to herbivores because their function limits the extent to which they can be mechanically protected. Virtually nothing is known about the fate of root production in Hong Kong. Presumably, soil invertebrates (such as the larvae of cicadas and chafers) consume a proportion of it and the rest is left for detritivores and microorganisms. Storage roots and rhizomes (underground stems), although a minor part of the total below-ground production, seem to make up much of the diet of the wild pig (*Sus scrofa*), judging by their digging activities, and are probably eaten also by the Chinese porcupine (*Hystrix brachyura*). Macaques sometimes eat the rhizomes of grasses, particularly the widely-sown exotics, *Paspalum notatum* and *Cynodon dactylon*.

The bulk of living plant material in streams consists of algal cells. Where macrophytes do occur, studies in other regions suggest that their tissues are rich in toxins and thus the plant cannot be eaten until these compounds have been leached out after death. However, snails and other invertebrates may graze algae which grow on the leaves and stems of aquatic plants, and so, incidentally, benefit the macrophyte by cleaning its photosynthetic surfaces. Algae encrusting rocks are the sole food of at least two benthic fishes in Hong Kong streams: *Liniparhomaloptera disparis* and *Pseudogastromyzon myersi* (Balitoridae). In Tai Po Kau Forest Stream, both fishes ingest large quantities of filamentous blue-green algae (*Homoeothrix*) and smaller amounts of diatoms and other algae; diatoms are a more significant dietary item for *Liniparhomaloptera* than for *Pseudogastromyzon*. The guts of these fishes contain small amounts of fine sand. The grinding action of these hard particles ruptures the mucilaginous sheaths enclosing *Homoeothrix* cells and so increases the efficiency with which the blue-green algae are digested.

Snails and a variety of mayfly larvae are major consumers of algae and the fine organic particles that accumulate on hard surfaces in freshwater habitats. This film of material is grazed more or less unselectively by a range of species, and the precise composition of the diet of these consumers changes among habitats reflecting the varying make up of the algal film. Among freshwater snails, however, *Physella acuta* (Physidae) (Fig. 5F) feeds selectively and consumes relatively large quantities of diatoms — a high-quality food. A comparison of movement patterns between this species and other, less-selective feeders, is instructive. *Physella* crawls rapidly in straight lines and searches a large area in a short time. By travelling straight ahead, it avoids revisiting sites grazed already and, by moving quickly, increases the chances of encountering rather rare patches of high-quality food. Snails with an unspecialized diet show less tendency towards the rapid, directed movement characteristic of *Physella*, but most species increase their rate of turning (i.e., show a tendency to move in circles) when a rich food patch is encountered. Such behaviour keeps the snail in the vicinity of the food until it is depleted, when the rate of turning declines and the snail crawls away.

Detritivores

Whatever the explanation for the low proportion consumed by herbivores, the bulk of primary production in terrestrial environments is left for the detritivores — organisms which consume dead organic matter. Dead organic material lying on the surface of the soil is termed litter. Estimates of the amount of litter present (comprising dead wood, bark, leaves, fruit, flowers and organic debris) vary between sites and with season, but dry weight values of 0.6 kg/m^2 for shrubland and over 0.8 kg/m^2 for forest give an indication of the range of values. The amount of leaf litter alone, however, is lower, with an average over the four seasons (summer, winter, autumn, spring) of around 0.2 kg/m^2 in forest. One study showed that annual litter fall amounts to just over 1.2 kg/m^2 in Hong Kong secondary forest, a figure which agrees well with estimates from climatically-similar regions elsewhere, but typhoons can increase the total quite substantially.

Unlike north-temperate latitudes (where over 80% of the annual litter fall occurs in autumn) there is no period in the year in Hong Kong when leaves are not being shed. This must have important implications for food availability to detritivores. Most local trees and shrubs are evergreen, retaining the old leaves until new foliage is produced in spring. One study in secondary woodland on Hong Kong Island recorded a litter-fall peak (approximately half of the annual total) during spring and early summer, which correlates with the main period of grazer activity (May to June). Herbivore attack on leaves can lead to their premature shedding, so increasing the magnitude of spring-summer litter fall, and it is noticeable that production of frass (faeces from grazing insects, especially caterpillars) peaks in May and June. A leaf which has lost a small percentage of its area to grazers is still a valuable photosynthesizing surface for the plant, but this is not the case where damage is excessive. If too large an area is lost, the respiratory costs of the leaf may exceed the photosynthetic gains.

Under these circumstances, the damaged leaf will be shed and enter the litter food chain. A secondary effect of this phenomenon is that few severely-damaged leaves remain attached to the plant, giving a false impression of the importance of herbivory.

The results from a single study of litter fall may not be applicable to the Territory as a whole because the seasonal pattern of leaf fall is affected by the species composition of the vegetation investigated. For example, at a site along the banks of Tai Po Kau Forest Stream there is a litter-fall peak from October through December, but this is contributed almost entirely by the litter of the deciduous tree *Liquidambar formosana* which happens to dominate parts of the area.

The major consumers of dead plant material in Hong Kong are invertebrates. They include a range of insects, such as the litter cockroach (*Opisthoplatia orientalis*) and its relatives, the litter stick insect (*Datames* sp.), many termite species (order Isoptera, of which 16 species are known from Hong Kong), and a host of smaller forms such as springtails (order Collembola), as well as a variety of other invertebrates. Among these are mites (order Acarina), snails (such as *Cyclophorus punctatus* and *Camaena* spp.), earthworms, woodlice (order Isopoda, several species in the families Armadillidae and Philosciidae), and amphipods (the exotic species *Talitroides topitotum*).

Although less selective than cytoplasm feeders, detritivores are not indiscriminate about what they eat. Newly-fallen leaves may still contain defensive chemicals evolved to dissuade grazing herbivores, and may not become palatable to detritivores until these compounds have been leached by rainfall. Microorganisms colonizing fresh litter will change its properties, softening it gradually and increasing the food value as protein-rich microbial tissues accumulate. In certain cases, detritivore feeding may be delayed until the leaves have been 'conditioned' by microbial action. The leaf species will also influence feeding by detritivores. For example, food preferences of four species of Hong Kong isopods (woodlice) reflect consumption of those species of leaf litter which can be assimilated most efficiently. However, isopods tend not to specialize exclusively on species of leaves yielding the highest energy gain, but vary the diet on occasion to incorporate litter with a high calcium or copper content. This reflects their specific need for copper (which is essential for synthesis of their blood pigment, haemocyanin) and calcium (a component of the exoskeleton). Significantly, isopods avoid eating litter of the camphor tree (*Cinnamomum camphora*) which contains a range of grazer feeding deterrents and retains its strong camphor smell long after it has been shed. Total isopod densities can exceed 500 individuals/m² (1.9 g/m²), and their feeding preferences will affect the fate of different species of fallen leaves. However, based on a knowledge of isopod standing stock and rates of feeding on a mixture of litter, they alone are likely to consume little more than 2% of the total annual leaf fall (equal to approximately 0.84 kg/m²/yr), although they may eat one third of the annual leaf fall of their preferred species.

Because shading and lack of nutrients limit algal growth and aquatic primary production, leaf litter is the major source of food in unpolluted steams. As on land, however, it is not consumed unselectively. There is evidence of strong preferences for particular leaf species by the detritivorous snail *Brotia hainanensis* (Fig. 5D), which eats litter of *Liquidambar formosana* much more readily than that of the wood-oil tree, *Vernicia* (formerly *Aleurites*) *montana*. This preference may be shared by other stream detritivores (mainly insects), which reach higher densities on submerged decomposing *L. formosana* leaves than on the tougher *V. montana* litter. *Bauhinia variegata*, which has particularly soft, palatable leaves, is also consumed with alacrity by stream detritivores, especially *B. hainanensis* and the shrimp, *Neocaridina serrata*.

In view of the large amounts of allochthonous detritus and the rapid leaf-litter processing in Hong Kong streams, it could be supposed that autochthonous primary production plays a minor role in animal nutrition. This supposition is only reasonable, however, if the algae and detritus have equivalent food value. There is reason to believe that this is not the case, and stream invertebrates may assimilate algae more efficiently than leaf litter. Much allochthonous detritus consists of structural polysaccharides (cellulose, hemicellulose and lignin) which are not readily digested by most aquatic insects, requiring colonization by microbes to make them palatable to consumers. Differences in nutrient content, tannin levels and leaf toughness among types of litter, as well as the species composition of colonizing microbes, will affect their consumption by invertebrates. By contrast, algal cells have a greater proportion of readily-assimilated material and therefore constitute a valuable food, although not all algae are similar in nutritional quality, digestibility, or the ease with which they can be removed from the substratum and ingested. Moreover, a simple measure of algal biomass may underestimate the importance of this food if the algal cells have a fast turnover. Additional difficulties will arise if invertebrates require mixed diets, eating detritus during one stage of their lives and algae in another.

As mentioned in Chapter 6, a measure of the relative importance of allochthonous and autochthonous food sources can be obtained by calculating the ratios between the rates of energy fixation by in-stream photosynthesis and the rates at which energy is consumed by the stream community. Any shortfall in energy demand must be made up from allochthonous sources. Measurements of this type in a shaded reach of Tai Po Kau Forest Stream have confirmed the dependence of the stream consumers upon allochthonous energy sources.

Breakdown of leaf litter — on land or in water — takes place through the combined action of leaching, physical processes (e.g., abrasion leading to breakage or fragmentation), microbial action, and detritivore feeding activities. These cannot be separated easily because, in nature, one does not occur in the absence of the others. One way of highlighting the role of detritivores is to enclose a known weight of leaf litter in mesh bags and expose them in the habitat of interest. Varying the size of the mesh used to make the bags will influence detritivore access and feeding activity. Retrieval of the bags, and reweighing the litter at various times after the initial exposure, provides data that can be used to calculate rates of litter breakdown (simply, the loss of mass of decomposing litter) in the presence or absence of detritivores.

Litter breaks down rapidly in Hong Kong forests, when compared to temperate regions, where fallen leaves may persist for a year or more. Leaves of *Ficus fistulosa* (the common yellow-stem fig) disappear from coarse-mesh (3 mm) bags rapidly: 2.3% of the initial weight is lost each day in summer, but the rate falls to 1.2% initial wt/day in winter. In fine-meshed (0.2 mm) bags, from which all animals but tiny mites and springtails are excluded, losses are about 0.6% initial wt/day in summer and less than 0.3% initial wt/day in winter. Evidently, litter breakdown proceeds more quickly in the warmer months when most rainfall occurs, and winter decomposition rates, while still considerable, are about half of those recorded in summer. Taking the annual average, rates of disappearance of litter from coarse-mesh bags (1.8% initial wt/day) are more than four times higher than from fine-meshed bags (0.4% initial wt/day), indicating the important role that larger detritivores play in litter breakdown.

Data from other local leaf species show that the rapid disappearance of *Ficus fistulosa* litter is not unusual. *Bauhinia purpurea* leaves in litter bags lose 2.3 to 4.7% of initial weight per day, the higher breakdown rate taking place during the warmer months. *Sterculia lanceolata* also shows high breakdown rates (3.9% initial wt/day), and this high value can be attributed to consumption of the litter (and litter bags!) by termites which were abundant in the secondary forest where measurements were made. Termites are confined almost entirely to the tropical region, with only a few species in the subtropics and none in the temperate zone. They live in colonies and this habit, combined with an ability to digest plant cell walls (with the aid of enzymes produced by mutualistic flagellate protozoans or bacteria in the hind gut), makes them effective and important consumers of leaf litter, bark and wood.

Litter breakdown in streams is rapid also — *Liquidambar formosana* and *Vernicia montana* leaves in Tai Po Kau Stream are reduced to tiny fragments within three months. Loss rates of 1.1% initial wt/day have been recorded for *Bauhinia variegata* in pools on an upper tributary of the Lam Tsuen River, although further downstream complete disappearance of leaves took little more than three weeks. From temperate streams, however, come many reports that leaves are not completely broken down after one year. The difference is probably attributable to higher water temperatures and hence faster rates of biological processing of litter in Hong Kong streams. Indeed, the few investigations of litter breakdown that have been undertaken elsewhere in the tropics indicate rapid disappearance of leaves, and summer versus winter comparisons in temperate regions indicate that elevated temperatures do speed decomposition.

To sum up, litter falls throughout the year in Hong Kong, although there are peak periods of leaf fall during spring and early summer. Despite some seasonal variation, litter breakdown in terrestrial and freshwater environments is rapid, probably because of high temperatures and, in terrestrial environments, the presence of termites and other detritivores which feed all year round. As a result, Hong Kong forests do not accumulate a deep or persistent litter layer; instead, fallen leaves are transformed rapidly into animal biomass through the activities of invertebrate consumers.

Coprophages

Herbivore dung is a special type of detritus that is utilized by an interesting community of animals known as coprophiles. Of these animals, the species which actually eat the dung are called coprophages, while the other species use the dung as habitat or prey upon the dung-feeders. Dung is chemically similar to plant litter because the digestible cell contents are removed in the herbivore gut leaving mainly cell walls in the faeces. It is physically quite different, however, and is consumed largely by dung specialists, at least in the initial stages. In Hong Kong, cattle are the primary providers of food and habitat for coprophiles. These animals include a wide range of species, although most are either flies (Diptera: Cyclorrhapha) or beetles (Coleoptera). The beetles are more diverse and include members of the families Hydrophilidae, Staphylinidae, Histeridae and Scarabeidae.

The scarab beetles include most of what are recognized conventionally as 'dung beetles', and their relatively large size makes them important contributors to the consumption and disappearance of dung. Newly-deposited dung is colonized rapidly by flying beetles; they tunnel into the soft material and feed actively with their specialized filtering mandibles which allow them to use only the soft fine-grained dung of the highest nutritive quality. Many dung beetles, especially small species (less than 1 cm long) in the genera *Onthophagus* and *Aphodius*, remain inside the dung pad, where they may mate and oviposit, but others (such as *Copris* spp. and large species of *Onthophagus*) tunnel into the soil beneath the pad and construct underground chambers where they store dung. The biggest dung beetles include those which shape balls of dung which are rolled away from the pad and stored in an underground burrow some distance away. Regrettably, these dung-rolling beetles (such as *Catharsius molossus*) have become rare in Hong Kong as cattle populations have declined.

The purpose of storing dung underground is to provide food for the beetle larvae which, unlike the adult beetles, have strong, hardened (sclerotized) mandibles and, with the aid of mutualistic cellulose-digesting bacteria in the gut, are capable of dealing with the plant fibres which constitute much of the dung. Beetles which do not make burrows oviposit in the dung pad where their larvae feed and grow, although they leave the pad eventually to pupate in the soil. Larvae of these species are found only during the cooler months when pads persist long enough to allow them to complete development. The larvae inhabit drier dung than the adult beetles which must leave the pad when it begins to dry out and they can no longer find easily-assimilated food or feed efficiently with their soft mandibles. As the pad dries further and is consumed by the dung beetle larvae, it comes to resemble plant litter more and more, and herbivore dung less and less. By the time the larvae have pupated, the pad consists of little more than dry plant fibres which enter the detritus food chain to be broken down by the action of microbes, termites, woodlice and so on.

Nectarivores and other flower visitors

Apart from leaves, stems and roots, the remaining fraction of the plant biomass consists of reproductive tissues — flowers, fruits and seeds. Here, the role of animal consumers can be very different from when herbivores eat non-reproductive tissues, and the plant-animal relationship is often at least potentially mutualistic. Although reproductive parts comprise only a small fraction of the total plant biomass, they provide food for many of the most conspicuous animals on local hillsides. Moreover, successful reproduction is essential for the persistence of a plant species within a community, so the influence of animals at this stage — both positive and negative — can be critical.

Except for a minority of wind-pollinated species (most conspicuously in Hong Kong — the grasses), plants depend on floral visitors for pollination. Potential pollinators visit flowers in expectation of a reward: usually nectar, pollen or both. Floral nectar functions solely as an attractant but is rarely freely available to all flower visitors. More often, the mechanical structure of the flower has evolved to restrict access, to a greater or lesser extent, so that potential pollinators are selected from the local array of nectarivores (nectar-feeders). Pollen may be an additional or sole reward but, unlike nectar, has a dual function as attractant and transmitter of genetic information. Flowers with no nectar reward typically produce a huge excess of pollen so the two functions do not compete. Like nectar, pollen may be freely available to all visitors or access may be restricted. For instance, in all local species of *Melastoma*, the anthers open by a small hole at the tip and pollen can only be extracted by the bee 'buzzing' the anthers at the correct frequency. There are no data from Hong Kong on what proportion of the available nectar and pollen is consumed by herbivores, but this is likely to be much higher than the proportion of vegetative tissues eaten, although the absolute amount is very much less.

Most flower visitors in Hong Kong are insects. Birds and bats are important pollinators in other tropical areas but are not of major significance here. However, the common white-eye (*Zosterops japonicus*) often visits flowers, and individuals with orange, pollen-covered heads have been mistaken occasionally for a different species! White-eyes have a brush-like tip to the tongue which may be an adaptation to drinking nectar. This bird may be the major pollinator of the red-flowered mangrove (*Bruguiera gymnorrhiza*) and it also visits cultivated specimens of the coral tree (*Erythrina* spp.) which are pollinated by birds elsewhere in Asia. In hillside shrublands, white-eyes can be seen visiting a variety of species, including the pink 'bells' of the Chinese New Year flower (*Enkianthus quinqueflorus*) and the large white flowers of the hillside shrub *Gordonia axillaris*. With *G. axillaris*, at least, the visits must be for insects as the flowers lack nectar and it is hard to see how a bird could gather pollen efficiently. Among other birds, the fork-tailed sunbird (*Aethopyga christinae*) and the two common flower-peckers (*Dicaeum* spp.) visit flowers in Hong Kong, but their ecological importance is not known. Mynahs (*Acridotheres* spp.) and starlings (*Sturnus* spp.) visit the large red flowers of cultivated *Bombax ceiba* but do not seem to visit native plant species.

Both local species of fruit bat (*Cynopterus sphinx* and *Rousettus leschenaulti*) are potential pollinators, but their importance as flower visitors in Hong Kong is not known and the local flora lacks most of the genera that are pollinated by these and related bats elsewhere. We suspect, however, that fruit bats pollinate the large, greenish flowers of the rose-apple (*Syzygium jambos*). This native of Malaysia is widely planted in Hong Kong and has run wild in some places. The numerous long stamens and abundant nectar are typical of flowers pollinated by vertebrates and, although white-eyes do visit this species, the pale colour, distinct fragrance and nocturnal opening of the flowers suggests that they are targeted at bats.

The major insect visitors to flowers are groups which are important world-wide: bees, wasps, butterflies and moths, hoverflies, and beetles. Honey bees (mostly the exotic honey bee *Apis mellifera*, but including native *A. cerana*) are the most abundant visitors to many plants but there are many other bee species in Hong Kong, including carpenter bees (*Xylocopa* spp.), blue-banded bees (*Anthophora* spp.), and leaf-cutter bees (*Megachile* spp.). Carpenter bees are the commonest visitors to the 'buzz-pollinated' *Melastoma* mentioned above, and have been shown to be the most effective pollinators of related species in Singapore. The same bees seem to be the only visitors to many large, bilaterally-symmetrical flowers, such as those of *Alpinia speciosa*. All such 'large-bee' flowers open during the warmer months when *Xylocopa* is active. The effectiveness of the smaller bee species as pollinators of these flowers is unknown. Butterflies are the major visitors to and presumed pollinators of a number of plant species in which the corolla forms a long tube, protecting the nectar from 'short-tongued' insects. Common examples of probable butterfly-pollinated flowers are *Jasminum lanceolarium, Pavetta hongkongensis, Pittosporum glabratum* and *Reevesia thyrsoidea*. Butterflies also visit unspecialized 'open' flowers, such as those of the Lauraceae. The hawk-moth (Sphingidae), which include both day-flying and night-flying species, visit many of the same species as butterflies. Members of the day-flying hawk-moth genus *Macroglossum*, which includes several local species, resemble tiny hummingbirds as they hover in the air, probing blooms for their nectar with a long, extensible proboscis. The resemblance to a small bird is heightened by a tuft of hairs at the end of the moth's abdomen which look like tail feathers.

Although the utilization of floral resources by insects may seem a perfect example of mutualism (a relationship in which both partners benefit), 'cheating' by one or other partner is common. Some flowers offer no reward and few give any external indication of when the reward is exhausted. Empty anthers of *Melastoma*, for instance, look just as attractive as full ones. Conversely, most flower visitors probably make little or no contribution to pollination, in effect, stealing the reward offered. One clear example of such theft is the habit of some *Xylocopa* bees of biting holes at the base of flowers with long corolla tubes to reach the otherwise inaccessible nectar.

The best example of a relationship between plants and their pollinators where both parties benefit is that between figs (*Ficus* spp.) and their wasp pollinators (Agaonidae). Each fig species has a different wasp species as pollinator and neither can survive without the other. The fruit-like figs (technically termed syconia) of the fig plant are, in fact, hollow receptacles with the numerous, unisexual flowers on the inside. These flowers are pollinated by tiny wasps which force their way into the syconium through the only opening (the ostiole) which is protected by overlapping scales. Once inside, the wasps pollinate the female flowers and lay their eggs in the ovaries of some of them. The wasp inserts its ovipositor down the style and can only reach the ovary in short-styled flowers. In monoecious fig species (i.e., those with both sexes on the same plant, such as *Ficus microcarpa*) each syconium contains female flowers with a range of style lengths. The ovaries of the long-styled flowers develop into seeds while wasp larvae develop within the ovaries of the short-styled flowers. These wasps emerge at the same time as the male flowers produce pollen. After mating, the young female wasps collect pollen and leave the syconium in search of young syconia at the 'female stage', where the cycle begins again.

In the dioecious fig species (i.e., those with separate male and female plants, such as *Ficus variegata*) the syconia on female trees contain only long-styled flowers, so the wasp cannot oviposit in the ovaries. These syconia produce only seeds. The syconia on male trees contain male flowers and short-styled female flowers. The ovaries of the short-styled flowers are usually occupied by wasp larvae so no seeds are produced and the plants are functionally male.

In both monoecious and dioecious fig species, the situation is further complicated by the presence of other species of wasps, from several related families, which compete for oviposition sites but do not act as pollinators. Most of these species oviposit from outside the syconium. Even if they carry pollen, the adult wasps never come into contact with the stigmas of the flowers, and hence do not contribute to pollination.

Frugivores and seed predators

As with pollen and nectar, the importance of fleshy fruits to the local fauna is much greater than their relatively small contribution to the total plant biomass would suggest. In most individuals of most plant species, the majority of the fruits are removed by birds or mammals before they can fall to the ground. Note that the term 'fruit' is used by ecologists in a functional sense, rather than in the traditional morphological sense. The fleshy tissues around the seed (the fruit pulp) may, from the strictly morphological point of view, be formed from fruit tissues, from the seed coat (aril or sarcotesta), or from other parts of the flower or inflorescence. Functionally these are all similar and, because they have evolved to attract animals, we would not expect them to possess mechanical or chemical defense. When defenses are present in the ripe fruit they, like the mechanical protection of pollen or nectar, may serve to target a restricted subset of the potential frugivores — the efficient dispersers. 'Efficient' in this case refers to an animal which removes the seed without damaging it *and* transports it to a site suitable for germination. Removal of the seed followed by deposition in an unsuitable site is as bad as no dispersal at all.

Potential dispersal agents visit the fruiting plant in expectation of a reward but fruit of high nutritional value does not seem to be essential to attract them. The pulp of many fruits is made up largely of water, and most are low in both calories and protein. The convenience of a food source that is displayed openly, is undefended and does not fight or run away, may make lower rewards acceptable. Moreover, it may be impossible for the frugivore fauna to learn which are the best fruits from among the diverse and continuously changing supply, thus allowing some plant species to 'cheat'. The extreme example of this is the leguminous tree genus, *Ormosia*, where the bright red seeds offer no pulp reward but must *look* edible to naïve birds. Ripe *Ormosia* seeds may remain on the plant for up to a year, suggesting that total cheating is not a particularly successful strategy! At the other end of the scale are fruits of *Sapium*, *Sterculia* and *Machilus*, for instance, where the pulp has a high fat content and is an excellent source of energy (Table 4).

Table 4 Nutritional content of the flesh of common Hong Kong fruits. The water content is expressed as the percentage of the fresh weight while the fat, soluble carbohydrate, and protein contents are the percentage of the dry weight.

Species	Water	Fat	Carbohydrate	Protein
Aporusa dioica	81	1	54	4
Ardisia quinquegona	86	2	55	7
Celtis sinensis	44	3	61	4
Diospyros morrisiana	69	1	51	4
Elaeocarpus sylvestris	65	2	52	5
Euonymus chinensis	52	43	5	11
Eurya nitida	70	1	70	4
Ficus fistulosa	84	1	71	2
Gnetum montanum	75	4	63	8
Ilex pubescens	77	1	60	5
Lantana camara	77	1	51	6
Lasianthus chinensis	93	2	88	4
Machilus thunbergii	64	41	19	8
Melastoma sanguineum	82	1	64	12
Psychotria rubra	91	2	70	5
Rhodomyrtus tomentosa	80	1	62	3
Sapium discolor	12	70	4	6
Schefflera octophylla	68	19	40	6
Sterculia lanceolata	62	31	20	17

The fruit pulp, which has evolved to be eaten, surrounds the seeds, which have not. Given their importance to the plant, we might expect the seeds to be the most heavily defended plant part of all. This seems often to be the case, although there is little direct evidence of this for local species. However, seeds are also of high nutritional value since they consist largely of food stores for the young seedling. The ripe fruit thus presents a paradox to fruit-eating animals: an edible envelope of typically low nutritional value surrounding one or more hard, toxic, but nutrient-rich seeds. Animals which eat fruits usually specialize on either fruit pulp or seeds, rarely both. This dichotomy in specialization is strikingly illustrated by the different ways in which bulbuls (*Pycnonotus* spp.) and the introduced Pallas's squirrel (*Callosciurus erythraeus*) treat fruits of the camphor tree. Bulbuls swallow the fruit whole but digest only the pulp, regurgitating the seed intact. Squirrels feeding at the same time in the same tree, strip off the flesh and discard it before consuming the seed.

The major fruit-pulp eaters in Hong Kong are birds, although the two fruit bat species mentioned above are frugivorous and fruit also forms a significant part of the diet of civets, macaques and probably other mammals, such as the ferret badger (*Melogale moschata*) and barking deer. Since most of these animals pass the seeds intact in their faeces (or regurgitate them) they are also major seed dispersal agents and hence influence patterns of vegetation succession (see Chapter 5). Many beetles (especially Bruchidae) attack seeds and can destroy the majority of a seed crop, however, few insects seem to eat fruit pulp in Hong Kong unless the crop ripens and falls to the ground. Exceptions to this generalization are species of owl moths (family Noctuidae) in the subfamily Ophiderinae, which pierce fruits and suck the juices using a proboscis tipped with minute teeth and spines. The hole made in the rind of the fruit provides an entry for fungi, and ophiderine moths are unpopular with citrus growers. Some curculionid beetles (weevils) also attack fruits, boring holes through the skin and pulp with the aid of a tiny pair of mandibles mounted at the end of a long snout. Eggs may be deposited in these holes and, after hatching, the grub-like larvae tunnel through the fruit as they feed. Seeds are attacked in the same way by some weevils.

Apart from insects, hillside rats and a number of bird species feed on seeds in Hong Kong. Unlike most seed-eating insects, which have evolved specialized detoxification mechanisms for particular chemical defenses and hence usually consume only one or a few related species of seeds, a relatively-large and long-lived vertebrate cannot afford to be so fussy. Both rats and birds, therefore, are probably restricted to seeds with limited chemical defenses. Grass seeds fit this description, with grasses apparently relying on their massive seed production to satiate seed-eaters so some seeds escape undamaged. The hillside rats probably eat some grass seed. Captive *Rattus sikkimensis* and *Niviventer fulvescens* extract and eat the seeds from some fleshy fruits but, with other species, consume the fruit pulp and leave the seeds.

Birds appear to be the major seed consumers. Flocks of the resident munias (*Lonchura* spp.) are most conspicuous as typically they take grass seed direct from the plant. The buntings (*Emberiza* spp.), in contrast, are winter visitors (and passage migrants) which seem to feed mainly on fallen seeds. Other major seed eaters in Hong Kong are the doves and pigeons, particularly the common spotted dove (*Streptopelia chinensis*), which is primarily a ground feeder. Elsewhere in the tropics, some species of pigeon are important seed dispersal agents but the local species apparently destroy in their crops most of the seeds they eat.

As well as grasses, other seeds likely to have reduced chemical defenses are those with mechanical protection. Several common tree and shrub species in Hong Kong produce hard, woody capsules which split at maturity to expose or release the seeds. Examples include the widespread bird-dispersed, pioneer tree (*Sapium discolor*) and two common species with winged, wind-dispersed seeds, *Gordonia axillaris* and *Schima superba*. In all three species, the capsules are attacked by squirrels before they split open naturally, and the seeds are consumed. Although Pallas's squirrel is an introduced species in Hong Kong there were presumably native squirrels with similar habits in the past.

The dry fruits and seeds of the Fagaceae (oaks, chestnuts, etc.) and a few other genera (e.g., *Camellia*, *Styrax*) have already been mentioned in the context of succession (see Chapter 5). The surplus seeds of these species serve as the only reward for their dispersers, so they cannot be heavily defended. Although such seeds are rarely, if ever, dispersed by members of the existing fauna in Hong Kong, they are eaten by rodents and, in some cases, macaques.

Carnivores and prey defensive adaptations

The problems of being a carnivore are very different from those of being a herbivore. Nutritional value is much less of a concern: in most cases the biochemical composition of the prey is similar to that of the predator, making it an excellent food. The major problem for predators is capturing sufficient prey, because most animals are not passive and immobile like plants, nor do they have to live in the open in order to photosynthesize. Like plants, however, chemical defenses do occur in some animals, particularly insects such as butterflies or moths and their larvae. Vertebrates may also use chemicals in defense: the unpleasant (to us) smell of shrews (especially the house shrew, *Suncus murinus*) seems to deter predators. It is surely significant that the house shrew, uniquely among small mammals, continually advertises its presence by a noise similar to the clinking of coins.

Toxicity in insects can result from the concentration and storage of poisons contained in the food plants. The poisonous nature of these insects may be advertised by bright colours: combinations of black, white and yellow, sometimes with the addition of red, are typical warning displays. The yellow and black pattern of the local tiger butterflies (*Danaus chrysippus* and *D. genutia*) is a good example. They and other members of the Danaidae, such as the crow butterflies (*Euploea* spp.), are toxic because their caterpillars consume poisonous milkweeds (Asclepiadaceae) and sequester the plant toxins in their tissues.

A distasteful or poisonous insect with bright warning colouration (termed an aposemete) obtains no advantage if predators (especially birds, which are visual hunters) do not recognize the advertisement. Indeed, the opposite may be the case because bright colours may attract unwelcome attention. Clearly, predators must learn to associate the bright colour with a distasteful prey, and the aposemetes are at risk during the learning process. The risk can be reduced, however, if different aposemetes have similar methods for advertising their 'unpleasantness'. It is probably for this reason that poisonous tiger butterflies, stinging vespid wasps, and toxic blister beetles (*Mylabris cinchorii*: Meloidae), from which the pharmaceutical cantharidin can be extracted, share the same black and yellow colouration. This similarity among unrelated aposemetes is known as Müllerian mimicry.

Batesian mimicry occurs when a harmless species (the mimic) has evolved to resemble an aposemete (the model), so that a predator is fooled by the resemblance and avoids the palatable mimic. The phenomenon is seen most commonly in butterfly mimics of danaid aposemetes, but certain hoverflies (Syrphidae) are also mimetic and are coloured so as to resemble bees. One local hawk-moth species, *Sataspes tagalica*, is an astonishing mimic of carpenter bees. The males are almost entirely black and, on the wing, closely resemble *Xylocopa phalothorax*. The female moths have yellow markings on the abdomen which give the impression of light reflecting off the shiny black body of *Xylocopa* bees; there is also a rarer, all-black female form of *S. tagalica*. Although the dynamics of Batesian mimicry have yet to be studied in detail for Hong Kong species, it is likely that the mimic must be significantly rarer than the aposemete, otherwise the predator will not learn to associate the unpleasant attributes of the model with conspicuous colours.

Local examples of Batesian mimicry by butterflies include the mimicry of poisonous *Atrophaneura aristolochiae* (Papilionidae) — the caterpillars of which feed on toxic *Aristolochia tagala* vines — by female *Papilio polytes* form *polytes*. The mimetic form of female *P. polytes* is less numerous that the 'normal' form (*P. polytes* form *mandane*) which resembles the male. As suggested above, the relative rarity of the mimetic form may be necessary to maintain — in the 'mind of the predator' — the association of a particular colour pattern with the unpleasant or toxic attributes of the model.

The mimicry of *Atrophaneura aristolochiae* by *Papilio polytes* form *polytes* involves two butterfly species in the same family. A more striking example involves males and females of the papilionid butterfly (*Chilasa clytia*) which occurs in two forms: *C. clytia* form *clytia* mimics the toxic danaid *Euploea core* while *C. clytia* form *dissimilis* mimics *Danaus similis*. Other mimetic relationships between butterfly families involve danaid models and (usually female) nymphalid mimics, including mimicry of *Danaus chrysippus* and the morphologically-similar *D. genutia* by *Argynnis hyperbius* and *Hypolimnas misippus*, of *Euploea midamus* by *Hypolimnas bolina*, of *Euploea core* by *Hypolimnas antilope*, and mimicry of *Danaus limniace* by *Hestina assimilis*. The females are more often mimetic than males, probably because deviation from the species' typical colour pattern by males results in a failure of females to recognize them as potential sexual partners, and a reduction in male fitness. This tendency towards conservation of the typical pattern in males is reinforced because female butterflies have the power of refusal in mating, and because males compete among themselves for mating opportunities.

An unusual example of possible Batesian mimicry is the resemblance to various tropical ant species achieved by members of the spider genus *Myrmarachne* (Salticidae). *Myrmarachne maxillosa* associates with the ant, *Polyrachis dives*, and may avoid bird predation by 'blending in' with groups of this aggressive ant which is unlikely to be a favoured dietary item because of a thick exoskeleton and an ability to defend itself by secreting formic acid when disturbed. Another *Myrmarachne* species (probably *M. plataleoides*) mimics the ferocious red tree ant *Oecophylla smaragdina*. The precise nature of the mimetic relationship between ants and spiders is unclear. *Myrmarachne* have been reported to prey upon their ant models, suggesting that the mimicry misleads the victims rather than, or as well as, the predators of spiders. However, observations made locally suggest that the spiders steal food from *Polyrachis* workers, suggesting that the relationship may be a combination of Batesian mimicry and kleptoparasitism (literally, 'food stealing'). Ant mimicry is also shown by the coreid (hemipteran) bug *Dulichius inflatus*, which closely resembles and associates with *Polyrachis* and has spines on the thorax similar to those of the ant.

In addition to defense by aposematism and mimicry, some beetles and hemipteran bugs produce unpleasant chemical secretions when threatened by predators. The pentatomid shield (or stink) bugs are familiar examples (especially green *Nezara viridula*), but water beetles such as surface-dwelling whirligig beetles (*Orechtochilus*: Gyrinidae) and the larger diving beetles (Dytiscidae such as *Cybister tripunctatus*), share this ability and are known to produce a number of defensive chemicals. Dytiscids secrete a range of compounds when squeezed or grasped by fish; some species produce toxic alkaloids, while others release steroids which anaesthetize fish temporarily. Water bugs (such as backswimmers: Notonectidae) also secrete noxious chemicals, and some of the larger bugs (*Diplonychus rusticum* and *Lethocerus indicus*: Belostomatidae) can inflict painful oral stings (bites) on humans.

Animals can also evade their predators by concealing themselves: crypsis is one way in which this can take place. A cryptic animal is coloured to match its normal background so that, although it is in full view of the predator, the predator does not distinguish the prey from its surroundings. There are many examples of cryptic insects in Hong Kong, coloured variously green, brown or pale brown, which are concealed when on a background of live vegetation or leaf litter. Grasshoppers and praying mantids are the preeminent examples, but many caterpillars, hemipteran bugs and spiders are cryptic also. Indeed, virtually every major faunal group contains at least a few species which are cryptic; consider, for example, the short-legged toad (*Megophrys brachykolos*) or Asian toad (*Bufo melanostictus*) sitting motionless among the fallen leaves on the forest floor, or a Chinese francolin (*Francolinus pintadeanus*) crouched among hillside grasses. Crypsis, often combined with a tendency to hide, is probably the primary method by which the majority of animals avoid detection by predators.

Terrestrial carnivores

Insects, both grazers and detritivores, consume much of the terrestrial primary production in Hong Kong and, in turn, provide the most important food source for carnivores, although most insectivores also eat other invertebrates. Insects are not a perfect food, even when not cryptic or poisonous, because of their indigestible chitinous exoskeleton. However, this does not seem to cause insectivores the problems that cellulose causes for herbivores, presumably because the chitin is largely on the outside of the insect and does not surround each cell.

Most inland bird species in Hong Kong are largely insectivorous although, as mentioned above, several of the commonest species eat large amounts of fruit. Even bulbuls and white-eyes, however, depend on insects during the early summer when fruit is scarce, and probably derive much of their protein intake from insects all year round. Their young require protein-rich insects, at least to start with, in order to sustain the rapid rate of growth which is typical of nestlings. There are many ways to be an insectivore and an immense diversity of species to eat, so the many insectivorous birds are not necessarily in direct competition. Swallows (*Hirundo* spp.), drongos (*Dicrurus* spp.) and flycatchers (various genera including *Muscicapa* and *Ficedula*) catch insects in flight, most warblers (such as *Phylloscopus* and *Prinia*) and white-eyes take their prey from leaves and twigs, while many thrushes (*Turdus* and *Myiophoneus*) search through the litter layer for their prey. Finer divisions are possible. For instance, aerial insectivores can be divided into 'screeners', like swallows and swifts (*Apus* spp.), which fly continuously for long periods, and 'sallyers', like the flycatchers, drongos and (particularly in spring) bulbuls, which make short flights from a perch. Some birds specialize on a particular type of insect: for example, the wryneck (*Jynx torquilla*) feeds principally on ants during the period of its migration that is spent in Hong Kong, using its long, sticky tongue to pick them up.

At night, bats take over the aerial insectivore niche while rats and shrews consume insects at ground level. Hong Kong has at least 19 species of insectivorous bats in the families Rhinolophidae (horseshoe bats), Hipposideridae (roundleaf bats), Molossidae (free-tailed bats), Emballonuridae (tomb bats) and Vespertilionidae (vesper bats). Moths and beetles — especially scarabeids, which are noisy and rather clumsy fliers — are the main prey, but representatives of all but the smallest groups of flying insects are taken and swarming adult termites are particularly favoured when available. The effect on prey populations may be considerable, because an insectivorous bat can eat the equivalent of one quarter of its body weight of insects in a single night. Little is known about the diet of hillside rats but two common species (*Niviventer fulvescens* and *Rattus sikkimensis*) favour insects when in captivity, although they will eat almost anything, and the remains of beetles and termites have been found in samples of their faeces. The diet of both native shrews consists of insects and other invertebrates. The house shrew is most abundant in the vicinity of human habitations although it has been caught in the countryside. The smaller grey shrew (*Crocidura attenuata*) is probably fairly abundant on hillsides although it is rarely seen and is difficult to trap. Shrews of the northern temperate zone (subfamily Soricinae) are renowned for their very high metabolic rates and voracious appetites. In contrast, shrews of the tropical subfamily, Crocidurinae, to which both local species belong, are reported to have lower metabolic rates and thus (presumably) lower food requirements.

Among Hong Kong's larger mammals, the Chinese pangolin (*Manis pentadactyla*) is entirely insectivorous. It uses its powerful forefeet to tear open ant and termite nests and its long, sticky tongue (like that of the wryneck, mentioned above) to pick them up. The pangolin has no teeth but the muscular stomach contains small stones to help grind up the insects after they have been swallowed intact. Civets also consume a lot of insects, particularly in early summer. Not surprisingly, they concentrate on the largest species, such as the litter cockroach (*Opisthoplatia orientalis*) and crickets (Gryllidae). Insects are probably also an important part of the diet of other, less studied, mammals, including mongooses (*Herpestes* spp.), the ferret badger and, perhaps, the leopard cat (*Felis bengalensis*). The stomachs of two road-killed ferret badgers contained the remains of earthworms, suggesting that these may be an important component of the diet. Lizards, frogs and toads (especially the abundant and widespread Asian toad, *Bufo melanostictus*) are also major consumers of insects and other invertebrates. Because of their commonness in a variety of habitats, skinks (several genera of Scincidae) and geckoes (*Gekko chinensis* and *Hemidactylus boweringi*) will be especially important in this regard, while the crested tree lizard (*Calotes versicolor*), which is mainly arboreal, may be a significant predator of shrubland insects.

Although vertebrate insectivores are more conspicuous, invertebrate predators probably consume far more insects in total. These include a huge (and largely unstudied) array of spiders, opiliones (harvestmen), centipedes and a variety of large insects, such as adult dragonflies, mantids, heteropteran bugs (Reduvidae and some coreids among them), tiger beetles (*Cicindela* spp. and *Tricondyla pulchripes*, which resembles a large ant), asilid (robber) flies, and many wasps (including some which are specialized hunters of other wasps), as well as many ant species which, through force of numbers, can subdue and kill prey many times larger than themselves.

Invertebrate predators and small vertebrates — whether herbivores or carnivores — in turn fall prey to larger carnivores. The top carnivore until relatively recently was the South China tiger (*Panthera tigris amoyensis*): the last substantiated sighting was in Shatin in 1947. The last tigers must have preyed largely on barking deer, wild pig and domestic cattle. Leopards (*Panthera pardus*) probably had a similar diet, although they generally take smaller prey than tigers when both species coexist and a choice is available. The leopard is also extinct locally, and our largest surviving mammalian carnivores are the leopard cat and two species of civet. The leopard cat is entirely carnivorous, and feeds largely on rats, birds and other small vertebrates. This is likely to have been true also of the large Indian civet (*Viverra zibetha*) which is probably extinct in Hong Kong. In contrast, the smaller civets — the palm civet (*Paguma larvata*) and the small Indian civet (*Viverricula indica*) — are real omnivores, eating fruit, insects, rats, birds and lizards. However, the two species of common hillside rat, *Niviventer fulvescens* and *Rattus sikkimensis*, make up the bulk of their diet.

Hillside rats, and the predominately lowland rat, *Rattus rattus flavipectus*, probably feature in the diet of many other vertebrate carnivores in Hong Kong, making them 'keystone species' in terrestrial communities. The two local species of mongoose (*Herpestes javanicus* and *H. urva*) and, probably, the ferret badger, eat rats also — although their relative contribution to the diet is not known — as do rat snakes (*Ptyas* spp., assigned to the genus *Coluber* by some experts), pit vipers (*Trimesurus* spp.), the Chinese cobra (*Naja naja atra*) and other large snakes. The biggest snake and the largest surviving vertebrate carnivore in Hong Kong is the Burmese python (*Python molurus bivittatus*), which can attain 6 m in length. Young individuals devour rats, but wild pigs, and perhaps barking deer, are likely to be the main prey of adults. There is at least one local record of a full-grown python swallowing a calf. While some snakes represent the final link in terrestrial food chains, others do not as shown by the incidence of ophiophagous (snake-eating) species in Hong Kong. Among them are the king cobra (*Ophiophagus hannah*) and two species of krait (*Bungarus* spp.) — all of which are highly poisonous — and the crested serpent eagle (*Spilornis cheela*) which, as its name suggests, specializes on snakes.

Hong Kong has a diverse fauna of birds of prey, although few of them are year-round residents. They have a wide range of diets and feeding techniques but, while most species show a degree of specialization, some are opportunistic and will eat almost any form of animal flesh, living or dead, when the opportunity presents itself. Owls are major predators on rats elsewhere but large owls are rare in Hong Kong and our commonest species, the diminutive collared scops owl (*Otus lempiji*), probably feeds mostly on insects such as large grasshoppers and crickets. The mainly nocturnal habits of hillside rats must also make them relatively safe from day-flying birds of prey, although they are sometimes taken by sparrowhawks (*Accipiter nisus*).

Birds dominate the diet of most falcons (*Falco* spp.) and the hawks (*Accipiter* spp.), although the kestrel (*Falco tinnunculus*), a common winter visitor, eats insects and other small prey such as lizards, typically dropping upon victims from a hover. The resident peregrine falcon (*F. peregrinus*) has long, pointed wings and is a bird of open country, where it swoops out of the sky on its hapless prey at incredible speed. Most hawks, in contrast, have short, rounded wings and are adapted for flying through forest, where they obtain much of their prey. The largest common species, the crested goshawk (*A. trivirgatus*), catches squirrels in addition to birds. The buzzard (*Buteo buteo*), another winter visitor, feeds on a wide variety of animals, from insects to vertebrates, dropping on its prey from a perch or while flying or hovering. As mentioned above, the crested serpent eagle eats snakes. The diet of most other eagles (*Aquila* and related genera) is elsewhere dominated by mammals but, because most of the surviving mammal fauna in Hong Kong is nocturnal, birds, reptiles and amphibians must be more important for the few eagles that are seen locally.

The common and conspicuous black kite (*Milvus migrans*), although capable of killing birds and other small vertebrates, is largely a scavenger. In Hong Kong it seems to feed mainly on dead fish and other edible waste floating in the sea. The consumption of carrion (recently dead animals) is not, strictly, carnivory, but is considered here because carrion forms part of the diet of many true carnivores. The most conspicuous sources of carrion in Hong Kong are the many animals killed on the roads. During the day, road-kills attract crows (*Corvus* spp.), magpies (*Pica pica*), the occasional black kite and probably other birds of prey, as well as semi-domestic and feral dogs. At night, fresh carrion is probably consumed opportunistically by owls (such as the collared scops owl) and mammalian carnivores (such as civets) as well as omnivores which do not otherwise eat vertebrate flesh, such as wild pigs, rats and, reputedly, even the Chinese porcupine.

Any animal flesh not consumed by vertebrates — as well as the small carcasses of toads, snails and so on — is exploited by insects. Several species of ants (including *Polyrachis dives*) are important consumers of small carcasses, and may exclude other insects from the food source. Where ants fail to dominate the carrion, it is colonized rapidly by fly maggots — a result of good 'carrion-finding' ability by sexually mature female flies. Sarcophagid flies (genera such as *Boettcherisca* and *Parasarcophaga*) increase the efficiency of colonization by depositing tiny larvae (larvipositing) directly onto the carcass. Calliphorid flies (blue-bottles, green-bottles and the like: *Lucillia*, *Chrysomya*, *Hemipyrellia*, etc.), in contrast, lay eggs around the orifices (nose, mouth and anus) of dead animals. Upon hatching, the larvae crawl inside the carcass and begin to feed. The large number of maggots that develop on a carcass compete for the limited food present because growth and development to the pupal stage must be completed before the remaining flesh is depleted. In certain flies, especially calliphorids, a shortage of food or extreme crowding stimulates starved larvae to pupate before growth is complete and give rise to undersized adults; some of these (in *Hemipyrellia ligurriens*, for example) may be as little as 10% of the potential adult weight. This strategy ensures that the maximum number of larvae survive to adulthood, and can be seen as 'compensation' for the delay in exploiting the food (due to egg-development time) when compared to larviparous sarcophagids.

Even when the flesh on the carcass is exhausted and the flies have gone, some nourishment remains. This provides sustenance for dermestid beetles which are able to digest the keratin which makes up the feathers and fur of birds and mammals. There are also other beetles associated with carrion: among them are staphylinids and histerids, which prey upon fly maggots, and sexton beetles (Silphidae). Silphids bury small carcasses as future food for their larvae, and the female guards the underground burial chamber while her offspring develop. Competition among the larvae is avoided by the female cannibalising some of her offspring if food remaining on the carcass becomes limiting.

The word 'probably' has featured in this account of the feeding habits of terrestrial carnivores a lot more than we would like. This reflects both the difficulty in studying wide-ranging, active animals and the absence of local researchers in this field. Only the civets have been studied in any detail and then only through the composition of their scats (faeces). In contrast, several species of freshwater carnivores have been studied in sufficient detail to merit a different approach — an attempt to go beyond 'What?' to 'Why?'

Foraging theory

For predators, some prey species will be relatively easy to capture, subdue and eat, while others will be more difficult; some prey will be widely distributed and common, while others may be highly localized or rare. The predator must therefore make at least two key decisions while it is looking for food: where are the best places to search? Which prey items (species) should be included in the diet? A major area of ecological research concerns foraging behaviour, which can be defined simply as the way in which animals acquire food and the choices that they make while doing so.

Predators — indeed, all animals — can be thought of as resource transformers, in that they ingest nutrients and energy in one form and transform them into body tissues, reproductive products or wastes. It will be apparent from Chapter 1 that their fitness and evolutionary success depends to a large degree on their efficiency at this process, in particular, the transformation of food into offspring. Where foragers, especially predators, can estimate and act upon information about food availability in their surroundings, they attempt to maximize their net rate of energy gain so maximizing the amount of energy that they have available for growth, reproduction, defense and so on. This makes good sense for an 'optimal animal', because fitness can be assumed to be an increasing function of the net rate of energy gain, and natural selection will favour animals which maximize this parameter. Note that the decision-making process need not be 'conscious', but those individuals with a tendency make the most appropriate decision, for whatever reason, will leave more offspring and (all other things being equal) will enjoy greater evolutionary success.

Where to eat? The world can be viewed as being divided into spatially-separate areas or patches; these may contain some or no food, but where food is present it will not necessarily occur in the same amounts. The foraging animal's problem is to determine in which sequence the patches should be visited, and the time spent feeding in each, so as to maximize energy gained and minimize time spent travelling between patches. An 'optimal' animal should exploit patches which yield energy at a high rate, and should abandon a patch when the net rate of energy gain falls to a level equivalent to the average yield from all of the patches in the habitat. In essence, a patch is abandoned or avoided when the predator would do better (on average) by feeding elsewhere.

What to eat? Different types of prey item are not equivalent, and an optimally-foraging predator must decide in what sequence food items should be eaten so as to maximize net energy gain, in the context of the costs of searching for, chasing, subduing and handling different prey items. On this basis, prey should be ranked and included in the diet in decreasing order of a measure of benefit (usually energy) derived from the prey divided by the costs of capturing, subduing and handling it. Accordingly, a prey item will be eaten when encountered if it will yield a higher-than-average net rate of energy gain compared to items in the habitat overall, but if the rewards are lower than the overall rate then prey should be passed by. In other words, prey are discarded or ignored if the predator would do better by obtaining another item.

Predictions about what and where to eat are derived from a large body of ecological research referred to as optimal foraging theory. We do not intend to review this material, but merely to suggest to the reader that animal behaviour and most adaptations (as discussed in Chapter 1) can be interpreted in the context of costs and benefits accruing to the individual. Usually, these benefits and costs are expressed in terms of energy, although a specific requirement for particular vitamins or minerals at certain stages in the life cycle may temporarily override energetic considerations. Two examples of analysis of Hong Kong predators using optimal foraging theory will be used to demonstrate the explanatory value of predictions based on measurements of costs and benefits. Both examples are freshwater invertebrates. They have been chosen because most foraging studies have involved animals such as birds which might be regarded as having sufficient 'brain power' to remember patch rankings and prey values, and to possess sufficient behavioural flexibility to act on this knowledge. The fact that some invertebrates forage optimally provides good support for the generality of these models.

Somanniathelphusa zanklon (Parathelphusidae) (Fig. 6B) is a freshwater crab found in lowland habitats in Hong Kong and South China, foraging in rice fields, irrigation ditches, flooded furrows, and slow-flowing streams and rivers. The crabs take a wide range of food, but readily eat freshwater snails when they are present. In the Lam Tsuen valley, five species of snail are potential prey for the crabs (Fig. 5): *Sinotaia quadrata* (Viviparidae), *Melanoides tuberculata* (Thiaridae), *Radix plicatulus* (Lymnaeidae), *Physella acuta* (Physidae) and the exotic *Biomphalaria straminea* (Planorbidae). A sixth species, *Brotia hainanensis* (Thiaridae), is found in the river but does not overlap in microhabitat with the crab and is not available as prey. Thiaridae and Viviparidae are heavy-shelled prosobranch snails, while the remaining three species are thin-shelled pulmonates which depend upon atmospheric air rather than dissolved oxygen.

In order to predict how crabs might choose prey among the snails available, we must first assign ranks to each species in terms of their food value. The most direct way of obtaining this value is to measure the energy gained from a particular snail divided by the costs of obtaining it. However, measurement of this benefit/cost ratio is impractical because of the difficulty of monitoring the energy expended by the crab while breaking the snail shell. Freshwater crabs attack snails by holding the shell in the chelae (pincers) and chipping away the lip of the shell aperture with the third maxillipeds (functionally equivalent to the jaws), or by holding the shell with the chelae and, starting with the apex, crushing the whorls with the maxillipeds. Both attack techniques take a significant amount of time, and the time taken to handle snails (from first grasping the shell to completing the meal) can be used as a convenient index of the costs of obtaining a meal. The benefits obtained from the prey item can be estimated simply from weighing the soft tissues of the snail. Prey value is thus tissue weight of a snail (benefit) divided by the time spent handling the snail (cost).

When individual crabs are presented with a single snail of known size, it is possible to measure the benefits and costs of each meal and thus to rank prey types. The crabs cannot crack *Sinotaia* shells, which are thick and heavy, and only relatively-large (two-year-old) individuals can break *Melanoides* shells. The lighter-shelled snail species can be eaten by large and small crabs, and the prey are ranked (in ascending order) as follows: *Melanoides* (large crabs only) < *Radix* < *Physella* < *Biomphalaria*. How do these measures of prey value affect the crab's choice of food when offered a mixture of snail species?

When given a mixed group of snails, the crabs show distinct feeding preferences. They seem to forage optimally, by choosing the most valuable prey (with the highest benefit/cost ratio) out of those offered. Thus *Biomphalaria* is eaten more often than the other species, and *Physella* is preferred over *Radix*. While it is satisfying to demonstrate that crabs seem to choose snail prey on the basis of a simple benefit/cost ratio, these results alone do not allow us to conclude that crabs forage optimally. It is possible that, for example, *Biomphalaria* is simply more 'tasty' than *Physella* which, in turn, is more palatable than *Radix*, and that selection is not based on economic considerations. It is difficult to measure how a snail tastes to a crab, but we can put our economic model to a further test. If the costs of handling some individuals of one snail species are increased, while the costs of handling other individuals of the same species are unchanged, we would expect members of the latter group to be preferred as prey if the crab is attempting to maximize net energy intake. This test can be undertaken by attaching a small plastic disc onto one surface of the flat, planispiral *Biomphalaria* shells using non-toxic waterproof glue. The plastic discs double the time taken by a crab to handle *Biomphalaria* shells, and reduce the benefit/cost ratio without influencing palatability. When given a choice between modified and unmodified snails, crabs strongly prefer normal *Biomphalaria*. This provides strong evidence that crab foraging behaviour is adjusted to maximize the net rate of energy gain.

It is reasonable to ask whether detailed observations on the foraging behaviour of individual predators can give us any insight into processes such as population regulation or the factors underlying community structure. It is notable that thin-shelled pulmonate snails, which are vulnerable to crab predation, are scarce in habitats where *Somanniathelphusa* is found, although heavy-shelled prosobranchs abound in such waters. By contrast, pulmonates build up dense populations in habitats where crabs are not present. These sites may lack crabs because they are polluted by agricultural wastes (which provide abundant food for snails) or, alternatively, because low dissolved-oxygen levels exclude the crabs but not air-breathing pulmonates. Nevertheless, it is clear that thin-shelled snails are abundant in the absence of their predators and rare in waters which contain crabs but otherwise appear suitable.

One curious aspect of the behaviour of snail-eating crabs is the tendency of the predator to eat only part of each snail prey when presented with several food items simultaneously, whereas all of the flesh is eaten if the snails are given to the crab one at a time. Apparently, when the crab is aware that many food items are present, it does not bother to finish one snail before starting on another. This partial prey consumption seems wasteful and, since the snail is discarded before all of the food value has been realized, it could be viewed as a case of a predator failing to maximize its net rate of energy gain. Appearances can be deceptive, however, as measurements of the feeding rate of the heteropteran bug *Diplonychus rusticum* (Belostomatidae) (Fig. 16), which is common in slow-flowing streams and marshes, have shown.

Heteropteran bugs have piercing mouthparts deployed to inject digestive enzymes into the prey after capture, whence the semi-digested 'soup' of tissues and body fluids is sucked out. They may spend a considerable time handling each item, and treat each prey as if it were

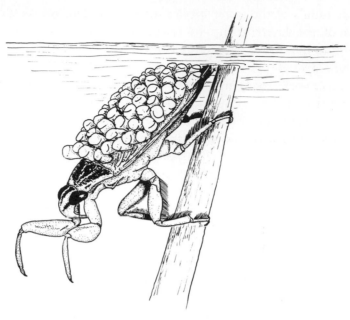

Fig. 16 A male *Diplonychus rusticum* (Heteroptera: Belostomatidae) carrying eggs. The female attaches her eggs to the wing covers of the male during egg-laying, where they remain for approximately 10 days until hatching. Drawing by David Dudgeon.

a resource patch. In this case, the rate of return from a single prey item declines as the time spent feeding on it increases, which is analogous to an increase in the difficulty of squeezing toothpaste from a tube as the tube empties — it is always difficult to get the last little bit out! Predators can maximize net energy gain at high prey densities by feeding for a short time on each prey item, and may abandon a meal while some extractable material still remains, provided that net energy intake can be enhanced by capturing more rewarding prey. On this basis, increased prey availability should reduce the time spent feeding on each item and result in partial consumption of prey.

Diplonychus feeds mainly on larvae of mosquitoes and chironomids (midges). When consuming a chironomid, the proportion of available food extracted increases with time, but the rate at which food is obtained declines over the same period. Thus *Diplonychus* obtains 33% of the extractable food after two minutes feeding, but after ten minutes of feeding only 60% has been extracted. At high prey densities, a predator which fed on two prey for a total time of four minutes would do as well as one feeding on a single chironomid for ten minutes. *Diplonychus* adjusts its behaviour so as to maximize energy gain: at high chironomid densities only 17% of the body weight of each larva is extracted before the prey is abandoned and another one attacked; at lower densities, approximately 70% of each chironomid may be eaten. Partial prey consumption by *Diplonychus* and the crab *Somanniathelphusa* is not 'wasteful'. Instead, it reflects a behavioural adjustment to maximize net rate of energy gain where prey are abundant.

Work on freshwater invertebrate predators suggests that simple decision rules underlie foraging behaviour, which is adjusted in such a way as to maximize the net rate of energy gain. However, we caution the reader that it is unreasonable to expect that all predators will conform to optimal-feeding models, because such models assume that animals have complete knowledge of their environment. The need to 'sample' novel prey items, or limits to sensory-processing capabilities and the ability to capture and eat certain prey, may be imposed by an animal's evolutionary history. These and other factors will restrict the degree of behavioural flexibility that is possible. Studies on predatory Odonata (dragonflies and damselflies) larvae from Hong Kong streams illustrate some of these limitations quite well.

The dragonfly *Zygonyx iris* (Libellulidae) (Fig. 7B) eats a subset of the potential invertebrate prey found in streams, but this is not a result of active selection. Instead, *Zygonyx* eats only those invertebrates which share its microhabitat of rock and boulder surfaces in torrential stream flows. Dietary composition in this instance is determined primarily by microhabitat overlap between predator and prey. Optimal foraging theory would predict that *Zygonyx* diets change as the predator grows and is able to ingest larger and more profitable prey, with the consequence that small prey are no longer consumed by large larvae. In fact, the range of prey eaten does increase as the predator grows and is able to ingest larger prey, but small food items are never dropped from the diet. In other words, this dragonfly eats anything it can handle, and does not ignore less profitable items. Larvae of the damselfly (*Euphaea decorata*) likewise add larger prey to the diet as they grow, while continuing to feed on small invertebrates. This behaviour is not suboptimal if the rate at which prey are encountered is low, because then the predator will always be hungry and will not be able to afford to let small, less-profitable items pass by.

Odonata larvae capture prey using an extensible hooked lower 'lip' (the labium) which is shot out from beneath the head to grasp prey as they swim or crawl past the predator waiting in ambush (Fig. 17); the unfortunate victim is then dragged back towards the

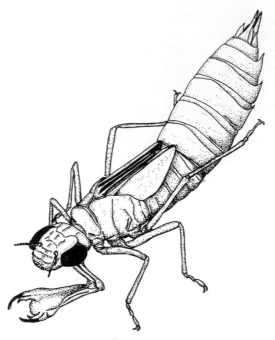

Fig. 17 An aeschnid dragonfly larva (*Anax* sp.) showing the extensible labium used in prey capture. Drawing by David Dudgeon.

waiting mandibles of the predator to be devoured. What can be included in the diet of dragonflies and damselflies depends on the size of the labium which is directly correlated with the size of the body. While certain invertebrate predators may exhibit flexible foraging behaviour, in the final analysis diet is constrained by evolutionary history manifested by the form of the feeding apparatus. Thus present-day behavioural interactions take place within well-defined phylogenetic limits.

This lengthy diversion into the details of predator foraging has been undertaken to illustrate the principles underlying food choice, and the way in which feeding behaviour can be altered as circumstances change. This is despite the fact that some predicted optimal behaviour might be constrained by the morphology or information-processing capabilities of animals. Given certain assumptions about predators, it is possible to make useful predictions about their behaviour based on optimal foraging models. The value of these models is that their success is not restricted to particular categories of habitats or types of animal; in principle they can be applied to terrestrial and aquatic habitats, both to vertebrates and invertebrates.

Carnivores in streams

While we may have some understanding of the factors underlying prey selection and feeding behaviour of some stream carnivores, our knowledge of the effects that predators might have on aquatic communities in Hong Kong is limited, although some remarks on the significance of predation by crabs have been made above. Studies on predatory fishes in Tai Po Kau Forest Stream have shown that a wide variety of invertebrates are eaten. The loach *Noemacheilus fasciolatus* and the goby *Ctenogobius duospilus* feed predominately on aquatic insects, especially chironomids and baetid mayfly larvae, and their diets are strikingly similar. The species of prey eaten are those insects which make up the bulk of the drift in Tai Po Kau Forest Stream, i.e., those animals which are regularly carried downstream by the current. Drifting behaviour is cyclic: most individuals and species drift at night which may reflect a need to avoid visually-hunting predators such as fish. The process can be thought of as an energetically-efficient way of searching for and colonizing new areas of the stream bed. Another view of drift is that it constitutes the 'doomed surplus' or animals which are excess of the carrying capacity of the stream bed; some of these may be weak or diseased individuals washed downstream to certain death. From this perspective, fishes are unlikely to influence benthic communities by predation because the prey taken would have been lost from the habitat one way or another.

The most straightforward way of estimating the effect of predatory fish on stream benthos is to enclose parts of the stream bed and add or remove fishes in the enclosure. In this experimental design it is important that the enclosure is made of mesh of an appropriate size that permits movement of invertebrate prey in and out of the cage but prevents the fish from so doing. One experiment of this type set up in the pools of a small hillstream and involving the predatory loach *Oreonectes platycephalus* (which is restricted to headwater streams) provided evidence that, at natural densities, fish do reduce the abundance of some benthic invertebrates. However, no influence on the relative abundance of major species was detectable. When the experiment was repeated in a riffle reach, predatory fish again had an effect, but the predation impact varied according to season, and was significantly greater during the dry season. Impacts were reduced during the wet season, when the stream bed was frequently disturbed by spates. Perhaps such spate-induced abiotic disturbances override the effects of biotic interactions during the wet season, with the result that processes such as predation and competition are more important as determinants of community structure during periods of stable stream discharge.

Predation of stream animals by terrestrial carnivores is one of the many links between terrestrial and aquatic communities. Various kingfishers, Chinese pond herons (*Ardeola bacchus*) and little green herons (*Butorides striatus*) take fish from shallow water. The risk from avian predators keeps larger fishes away from the shallow water close to the stream banks, which may force smaller fish into the shallows where they are less vulnerable to cannibalism or attack by predatory fishes. Small fishes run less risk of attack from birds because, for the bird, the return from a small fish may not be worth the effort expended in catching it. Where large fish are unavailable to birds, however, small fish in shallow water are at risk, and their behaviour can be seen as a balancing act to reduce the chances of encountering predatory birds or large fish. Note that if fish behaviour is affected by predatory birds, and if the fish have an impact on populations of benthic invertebrates, then by influencing fish behaviour the birds indirectly affect invertebrate populations.

Emerging aquatic insects fall victim to terrestrial invertebrate predators and birds including the plumbeous water redstart (*Rhyacornis fuliginosus*) which forages along shaded streams. Brown dippers (*Cinclus pallasi*) plunge into the water to capture larval insects, although these birds are very rare in Hong Kong. Wagtails, especially the grey wagtail (*Motacilla cinerea*), patrol stream banks searching for stranded insects along the shoreline; both drowned terrestrial insects and adult aquatic insects are eaten. Daubenton's bat (*Myotis daubentonii*) forages for insects over streams soon after nightfall, while Rickett's big-footed bat (*Myotis ricketti*) captures fish resting close to the surface using its large feet which have well-developed, hooked claws. One consequence of the large feet is that *M. ricketti* is unable to groom itself efficiently or keep the fur free of parasites. The evolutionary response by this bat has been a partial loss of hair, which reduces the chance of external parasites (mites, ticks and hippoboscid flies) gaining a hold and establishing themselves.

Other predatory vertebrates found in and around Hong Kong streams are water snakes (*Opisthotropis* spp., *Enhydris* spp.), the semi-aquatic waterside skink (*Tropidophorum sinicus*), terrapins (e.g., *Cuora trifasciata*, *Clemnys bealei* and *Platysternon megacephalum*), frogs (e.g., *Amolops hongkongensis*, *Rana livida*, *R. exilispinosa*, *R. spinosa*), and the Hong Kong newt (*Paramesotriton hongkongensis*). At one time this list would have been completed by three larger carnivores. The eastern Chinese otter (*Lutra lutra chinensis*) was thought to be extinct locally, but there have been a small number of recent sightings of this species at Mai Po Marshes. The brown fish owl (*Ketupa zeylonicus*) was said by ornithologist Robert Swinhoe to be 'pretty abundant' on Hong Kong Island in 1860 and, although G.A.C. Herklots reported that they bred at Pokfulam until 1953, the species is now very rare. This owl was associated with watercourses and spent much of the day resting on the ground concealed among boulders. Swinhoe noted that the diet of the brown fish owl (based on examination of regurgitated pellets) consisted chiefly of freshwater crabs and some fish. The water monitor (*Varanus salvator*) is almost certainly extinct, but would once have been common along rivers and streams. This lizard can reach 250 cm in length. The last records of wild specimens in the New Territories date from 30 years ago and the most recent record from Hong Kong Island was in 1930.

Any account of freshwater predators would be incomplete without mention of what were undoubtedly the largest of these animals — the crocodiles. Skeletal remains of the fish-eating long-nosed crocodile or false gharial (*Tomistomus schlegeli*) are known from the Pearl River, but similar bone fragments uncovered locally cannot be identified with certainty. Whether the much larger brackish-water crocodile, *Crocodylus porosus*, which once inhabited the mangroves of South China and is a known man-eater, ever penetrated Hong Kong's freshwaters is not certain, but the possibility that these magnificent reptiles once lurked in the Territory's rivers is an alluring one.

8

Aliens

When people first started clearing patches of forest for cultivation and settlement, an entirely new type of habitat for plants and animals was created. The most distinctive characteristic of this new habitat was that it combined high light intensities with adequate water and nutrients. Tree-fall gaps in forest have these characteristics but they are usually small and always short-lived, with regeneration dominated by woody species. Before the arrival of people, permanent open habitats were confined largely to sites such as cliffs and beaches, where soil conditions prevented the formation of a closed woody canopy, and eroding river-banks, where disturbance kept the site open.

At first, the newly-cleared habitat would have been occupied by plants and animals from the surrounding natural habitats. Previously-rare inhabitants of cliffs, riverbanks and other open areas expanded into the man-made sites to which they were at least partly pre-adapted. Intense natural selection probably led to rapid evolution of distinct, 'weedy' varieties, better adapted to living with man. As human impact spread and trading links developed, the plant and animal weeds of different areas also spread and mixed. The regional floras and faunas of man-made open habitats became increasingly homogenized until, from the sixteenth century onwards, the process of globalization began. The ultimate limits to this process are set not by natural dispersal abilities, but by the climatic tolerance of the species concerned.

Globalization has proceeded at different rates in different groups of organisms. It has been most dramatic with plant weeds, where now even a trained botanist would find it hard to identify what continent he or she was on from the plant life of a tropical city. For example, 500 years ago Hong Kong and the island of Dominica, in the Caribbean, probably had no inland plant species in common; today they share more than a hundred weeds of man-made open habitats. Invertebrates have also participated in the globalization of the earth's biota, because many smaller forms are carried undetected among other goods and so have dispersed over the planet to wherever man has spread. Indeed, many stored-products pests and insects associated with widely-cultivated crops have almost cosmopolitan distribution patterns. In contrast, there have been very few intercontinental exchanges of birds and mammals, although several species of freshwater fishes have become quite widespread during the last century.

Plants

Hong Kong's total vascular plant flora of approximately 2000 species includes at least 145 naturalized aliens or exotics: that is, species introduced from other parts of the world which have run wild in Hong Kong. Of these, slightly over half were probably introduced from tropical America (South and Central America, and the Caribbean), 18% from northern Eurasia (including Europe), 15% from Africa or Madagascar, and 10% from tropical and subtropical Asia. The apparently small proportion from Asia is misleading, however, as weeds of Asian origin, unless they became established recently, would not usually be recognized as introductions. Hong Kong's flora includes many weedy species, such as *Emilia sonchifolia* and *Vernonia cinerea*, which are undoubtedly of Asian origin but are unlikely to have found a suitable habitat in the primeval, forested landscape. Moreover, there are a number of species, such as *Cynodon dactylon*, which spread throughout the tropics so early in man's recent expansion that their region of origin is obscure and they are treated as natives. The real figure for naturalized exotics in Hong Kong should probably be 250 to 300 species.

The exotic flora must be able to tolerate local climatic conditions in the same way as the natives. The presence of at least 26 species which have apparently been introduced from temperate northern Eurasia — most probably, Europe and the Mediterranean — is an indication of the transitional nature of Hong Kong's climate. In contrast, only one of these Eurasian species (*Plantago major*) has successfully established itself in equatorial Singapore. Elsewhere in the tropics, aliens of temperate Eurasian origin are usually confined to high altitudes where mean temperatures are less than those in the lowlands. In Hong Kong, many of these species grow most actively in the cooler, winter months.

It is more difficult to identify species with the opposite distribution, that is, widespread in tropical lowlands but excluded from Hong Kong by cold. The 60 or so species naturalized in Singapore but absent from Hong Kong could be explained by chance or climatic factors other than temperature, such as Hong Kong's more seasonal rainfall. That winter cold could be a factor, however, is shown by the damage to many species of tropical origin caused by the unusually cold weather of 28 December 1991 (see Chapter 3). The most sensitive species was *Mikania micrantha*, a creeper of South American origin, which was severely damaged by temperatures in the range 3–5°C. Several other species of tropical origin suffered chilling injury and the above-ground portions of all those found above the frost line (around 400 m) on Tai Mo Shan were killed. In contrast, no species of temperate origin was damaged by chilling temperatures and some, such as the dandelion (*Taraxacum officinale*), were undamaged on the summit of Tai Mo Shan, where temperatures fell to around −5°C.

For most exotic species, the means of introduction to Hong Kong is not definitely known. At least 55 species of the current exotic flora were recorded as already naturalized on Hong Kong Island in George Bentham's *Flora Hongkongensis*, published in 1861, and some could have been in the region for centuries. Significantly, the biologist Berthold Seeman, who visited the island in 1850, recorded the now-naturalized *Lantana camara*, *Mimosa pudica* and *Passiflora foetida* only in cultivation. That *Lantana* has escaped at least three times is shown by the presence of at least three, distinct, naturalized cultivars in Hong Kong. At least another 50 species are probably escapes from cultivation, including ornamentals (e.g., *Oxalis corymbosa* and *Thunbergia alata*) and crop plants (e.g., the fodder grasses, *Brachiaria mutica* and *Panicum maximum*). The remainder were most likely introduced accidentally, many probably as seeds in the soil which accompanied species imported deliberately. It is important to note, however, that the great majority of the thousands of plant species introduced deliberately or accidentally to Hong Kong have never become naturalized. This includes species such as the widely-planted paper-bark tree (*Melaleuca quinquenervia*) which have become aggressive weeds elsewhere in the tropics.

The term 'naturalized' suggests becoming part of the native flora but, in fact, most exotic plants remain confined to recently-disturbed sites, such as urban wasteland, cultivated areas and newly-abandoned cultivation. Except along major paths, recognizably exotic species have not invaded hillside grassland, shrubland or woodland. Indeed, the presence of persistent exotic species such as *Lantana* is a good marker of abandoned settlements. The only place where exotic plants survive in the absence of continued human disturbance is along the coast, especially at the back of sandy beaches, where prickly pear (*Opuntia stricta*), *Rhynchelytrum repens* and *Tridax procumbens* have become a permanent part of the flora. How exotics are excluded from other vegetation types is far from obvious. The simplest explanation is that the process of introduction favours species which can persist only at sites with high light and nutrient levels and low competition. In other words, cultivated plants and weeds of cultivation are likely to be the most successful exotics, but these plants cannot compete in closed vegetation or on poor soils. It does seem odd, however, that out of the large number of introductions apparently none have the adaptations to compete in native secondary vegetation.

Birds

Identifying exotics in Hong Kong's bird fauna is even more difficult than with plants. The popularity of cage birds in Hong Kong results in thousands of accidental escapes each year as well as deliberate releases. Enough budgerigars (*Melopsittacus undulatus*) or yellow-fronted canaries (*Serinus mozambicus*), for instance, must escape every year to found a wild population if the Hong Kong environment was at all suitable. Yet only a few species with apparently established wild populations, such as the lesser sulphur-crested cockatoo (*Cacatua sulphurea*), are of such recent introduction and so obviously outside their native range that they can be spotted immediately. A number of introduced species, such as the rainbow lorikeet (*Trichoglossus haematodus*), have bred in the wild but are not well-established, while the population of the azure-winged magpie (*Cyanopica cyana*), which was founded by four escaped birds in the Zoological and Botanical Gardens, expanded for several years before declining subsequently. Because many cage birds originate in South China, distinguishing escapes from spontaneous range extensions is no easy matter.

On the other hand, only a minority of our supposedly native bird fauna is adapted to life in forest, which, until it was cleared by man several centuries ago, would have been the most extensive natural habitat in the Territory. Tropical countries at an earlier stage of deforestation, such as Malaysia today, have very distinct communities of birds in forest and open country, with hardly any species in common. There, the open-country bird fauna consists of native coastal species, originally confined to mangroves, beach scrub and river estuaries, together with exotic species which have either escaped from captivity or extended their range as deforestation progressed. In Singapore, where forest is confined to a few small patches, most of the bird fauna of the forest interior is extinct, but open-country birds are still confined to the forest edges and the upper canopy. As we have seen, Hong Kong must have been almost completely deforested in the recent past and open-country species now dominate the bird fauna. Many, perhaps most, of our common hillside birds have probably extended their range gradually into man-made open habitats as the deforestation of South China progressed. In the strictest sense, they are exotics, although this is usually impossible to demonstrate for any individual species.

We do have suggestive evidence, however, for one of the commonest open-country birds, the crested bulbul (*Pycnonotus jocosus*). Ornithologist Robert Swinhoe, who visited Hong Kong in 1860, stated quite clearly that this species was not present at that time, although it was abundant in Guangzhou. Swinhoe was not in Hong Kong long enough to make a comprehensive bird list but he would surely not have missed a bird he knew well if it was at all widespread. This bulbul has become naturalized from escaped cage birds or deliberate releases in many parts of the world, including Sydney, Mauritius, Hawaii and Florida, so a spontaneous or human-assisted range extension in deforested South China is not at all improbable. It is the commonest bulbul of the urban fringes in Hong Kong and is completely absent from forest: a distribution which is again suggestive of an introduced species.

Interestingly, the diversity of our forest-adapted bird fauna has risen over the past few decades as the area and maturity of plantations and secondary woodlands has increased. Many of the common resident birds of the forest interior at Tai Po Kau (the largest woodland area and the one most visited by bird watchers) were absent or only known as winter visitors ten to 20 years ago. The present populations of the scarlet and grey-throated minivets (*Pericrocotus flammeus* and *P. solaris*) and the chestnut bulbul (*Hypsipetes castanotus*) were probably established by winter visitors which stayed to breed. Birds of prey such as the crested goshawk (*Accipiter trivirgatus*) and the black bazza (*Aviceda leuphotes*) have also become more numerous and widespread in recent years, as has the barred owlet (*Glaucidium cuculoides*). Other birds now established in Tai Po Kau Forest, such as the velvet-fronted nuthatch (*Sitta frontalis*) and silver-eared mesia (*Leiothrix argentauris*), may be escaped or deliberately released cage-birds. Most of these 'new' species may have been part of Hong Kong's original forest bird fauna so they cannot be considered exotic. However, their successful (re)introduction raises an interesting question: since the bird fauna of Hong Kong's expanding forest area is clearly undersaturated by comparison with forests in South China, could we and should we (re)introduce more species? This question is considered further in the final chapter.

Other terrestrial vertebrates

Fewer mammals than birds are kept as pets so there are fewer escapes: still fewer survive. The clearest example of a naturalized exotic among the local mammal fauna is Pallas's squirrel (*Callosciurus erythraeus*), which became established in Hong Kong, presumably from escaped pets, some time in the late 1960s. Originally confined to a few sites, squirrels are now widespread on Hong Kong Island and in much of the New Territories, although apparently still absent from Lantau and the other islands. It appears that different subspecies have become established either side of Victoria Harbour, as a result of at least two separate introductions of squirrels from different parts of the species range. The correct scientific names for these subspecies are still in doubt but the Hong Kong Island squirrels closely resemble some populations in Thailand (*Callosciurus erythraeus thai*), while the New Territories squirrels seem to belong to *Callosciurus erythraeus styani*, a subspecies from Zhejiang Province and adjacent parts of eastern China. If these identifications are correct, the two populations probably originated more than 2000 km apart in areas with very different climates and natural vegetation. We might therefore expect considerable differences in their ecology, although no comparative studies have been undertaken. There must have been at least one *Callosciurus* species in Hong Kong's original forest fauna so, even though the two subspecies of Pallas's squirrel are exotic, they probably occupy a 'natural' niche.

However, both subspecies are noticeably much less common away from human habitations, suggesting that they are not yet fully naturalized.

By contrast with the success of *Callosciurus* introductions, Siberian chipmunks (*Tamias sibiricus*), which are sold widely as pets, have yet to become established in the wild although individual escapees are sighted occasionally. Perhaps this is because Hong Kong is outside the latitudinal range of this Palaearctic species, although this explanation does not seem entirely satisfactory as captive Siberian chipmunks thrive and breed locally. Similarly, hundreds of golden hamsters (*Mesocricetus auratus*) must escape every year and can survive for a short time, but the Hong Kong environment is obviously very unlike that of their native Syrian steppes.

The macaques found in the Kowloon Hills and elsewhere in the New Territories are an even more complex case than the squirrels. The rhesus macaque (*Macaca mulatta*) is native to the Hong Kong region but was not recorded in the wild in the first half of the this century although it was reported from the Lema Islands by nineteenth-century visitors. The other well-established species, the long-tailed macaque (*M. fascicularis*), is a definite exotic from Southeast Asia, and must have originated from released pets. The same is true for two other species present in small numbers in Hong Kong: the Japanese macaque (*M. fuscata*) and the Tibetan stump-tailed macaque (*M. thibetana*). Naturalist J.D. Romer was the first to report long-tailed macaques in Hong Kong after sighting a breeding group at Kowloon Reservoir in 1966; presumably the initial release (or escape) was some time before this record. If other macaque species were released — probably when cute baby monkeys became bad-tempered adults — we can be sure that rhesus were released too. We cannot be certain, however, that *none* of the current population are descended from survivors of the original Hong Kong macaques.

An interesting feature of the Hong Kong macaque population is the extent of hybridization. The earliest records and photographs of the long-tailed macaques at Kowloon Reservoir indicate that the breeding group of this species was initially monospecific. A recent study undertaken by John Fellowes for the World Wide Fund for Nature (Hong Kong), in contrast, found only one percent pure long-tailed macaques in the present population of the same area. Apparently pure rhesus macaques form about half of the population and most of the rest seem to be rhesus-long-tail hybrids. There are also several probable rhesus-Japanese macaque hybrids and possibly other combinations. It would be interesting to know if the predominance of rhesus macaques reflects a larger founder population (i.e., more releases or survivors) or superior adaptation to an environment which is within the native range of the species.

In 1992 there were around 800 macaques in Hong Kong, with over 700 of these in the Kowloon Hills area (in Kam Shan and Lion Rock Country Parks) and most of the rest in Shing Mun Country Park and Tai Po Kau Nature Reserve. The majority of macaques depend on feeding by people for much of their diet, although natural foods are eaten when available. Close contact with people results in scratches, bites and the risk of more serious injury and disease transmission. The macaque population appears to be increasing by about 10% each year, so these problems are likely to get worse. The best approach to controlling both aggressive interactions with people and long-term population growth is probably to reduce or eliminate feeding. However, there are considerable practical difficulties in stopping people feeding monkeys.

Among the small mammals, it is likely that all the species that are closely associated with human habitations arrived in Hong Kong with or after man: the house shrew (*Suncus murinus*), the house rat (*Rattus rattus rattus*), the Norway rat (*Rattus norvegicus*), and the house mouse (*Mus musculus castaneus*). The extent to which these species are found away from human habitations varies. The house shrew has been captured in natural vegetation in several places while the Norway rat has so far been found on hillsides only on Lantau Island. Among larger mammals, feral domestic cats and dogs occur widely in the New Territories and may have a large (although, as yet, unquantified) effect on the local fauna.

There are no definite exotics among the reptiles and amphibians of upland areas but, applying the same arguments as used above, it is likely that species now wholly dependent on man-made habitats were either introduced or extended their ranges with man. The four-clawed gecko (*Gehyra mutilata*) may be an exotic as it is not widespread in the Territory and largely confined to buildings. However, Hong Kong is within the natural range of this species. Brooke's gecko (*Hemidactylus brookii*), which is known only from two urban sites, is certainly exotic; Bowering's gecko (*Hemidactylus boweringi*), by contrast, is the common house gecko in Hong Kong, but occurs also in a wide range of other habitats and is presumably native.

Terrestrial invertebrates

Hong Kong's terrestrial invertebrates are too little known for exotics to be identified consistently. Most of the definite introductions, such as the familiar American cockroach (*Periplaneta americana*) and its allies, as well as certain crop pests, are largely confined to the vicinity of human habitation or agricultural land, but some have become more widespread. The terrestrial amphipod crustacean, *Talitroides topitotum*, has been spread from Australia across the tropics and the warm-temperate zone along with soil transported with ornamental plants. It is abundant in wooded habitats in the Territory (mostly on Hong Kong Island), where the soil is not too acidic and remains damp throughout the year. The giant African land snail, *Achatina fulica*, was first recorded in South China in 1931 and arrived in Hong Kong in 1941. It is now widespread and abundant; indeed, in the period June to October 1946, 74 746 *Achatina*, weighing approximately 1993 kg, were collected from the Botanic Gardens. Eggs of the snail were believed to have been brought to China from Singapore (where the snail is an agricultural pest) with the possible aim of using *Achatina* as a source of food. The snails originated in East Africa, but were introduced to India in about 1860 (via Mauritius) and, by 1910, *Achatina* was reported to be widespread there and in Sri Lanka; by 1922 it had become well established in Singapore and Malaysia. While *Achatina* has colonized a range of terrestrial habitats in Hong Kong, other introduced terrestrial molluscs, such as the slug *Incilaria bilineata* and the snail *Bradybaena similaris*, tend to be associated with human habitation, the former species being a minor agricultural pest.

One exceptionally successful introduction is the honey bee (or Western hive bee), *Apis mellifera*, which originates in Europe. This bee appears to have largely displaced the native Asian (or Eastern) hive bee (*Apis cerana*) and it is not clear to what extent the indigenous bee remains a significant pollinator of the local flora. Most other known invertebrate introductions (mostly insects) are associated with buildings, and include various moths and beetles which are pests of stored products, as well as the ubiquitous pharaoh ant (*Monomorium pharaonis*) and the tiny sugar ants (*Iridomyrmex anceps* and *Tapinoma melanocephalum*).

Undoubtedly the most damaging invertebrate introduction — indeed, the most damaging exotic species established in Hong Kong — is the pinewood nematode (*Bursaphelenchus xylophilus*). This nematode worm is probably native to North America, where it occurs in a number of native pine species without doing much damage. By contrast, in other parts of the world (where it was probably spread in infected pine logs) it causes the lethal pine wilt disease. The pinewood nematode was first identified in Hong Kong in 1982, although it must have been present considerably earlier. It is spread by longicorn beetles (Cerambycidae, particularly *Monochamus alternatus*); larvae of the beetles bore in pine wood while the adult beetles feed on the bark. Nematodes are carried in the tracheae (respiratory system) of infective *Monochamus* adults, and actively migrate to the feeding site damaged by the beetle where the worms penetrate the tissue of the pine host. Infestation of the beetle by the nematode involves the worms aggregating around the maturing pupae, and invasion of the tracheae of the emerging adult as it leaves the pine wood for a nuptial flight.

The nematodes can kill a pine tree within six months although some survive much longer. Native *Pinus massoniana* is extremely susceptible to nematode attack and the deaths of vast numbers of this pine have transformed the upland landscape. The exotic pines planted in Hong Kong (principally the slash pine, *Pinus elliottii*) are less susceptible. It should also be mentioned that there is evidence from Japan suggesting that air pollution can greatly increase the susceptibility of pine trees to nematode infection.

Aliens in streams

The most obvious alien species in Hong Kong streams are fishes. The guppy (*Poecilia reticulata*) and the mosquito fish (*Gambusia affinis*) originate from Central America but have pan-tropical distributions arising from deliberate introductions as part of mosquito-control programmes. Other exotics include escapes from the aquarium trade (a legacy of the days when it was a widespread practice to rear ornamental fishes for sale in outdoor ponds), of which only the swordtail (*Xiphophorus helleri*) and the platy (*X. variatus*) have established populations that are self-sustaining in the long term. They are in the same family (Poeciliidae) as the guppy and mosquito fish and, like them, originate from Central America. The success of poeciliids as invaders can be attributed, in part, to the fact that they do not lay eggs but instead are viviparous, producing well-formed young. Males fertilize the females using a gonopodium (a rod-like modification of the anal fin) which serves as a penis. The eggs develop and hatch inside the body of the female and the free-swimming young are released only after the yolk sac has been absorbed fully. There are two advantages in this breeding strategy: firstly, eggs and vulnerable fish larvae are protected from predation; and, secondly, the mobile female can respond to deteriorating environmental condition by movement, carrying the eggs to safety with her. Female poeciliids store sperm and can produce successive broods of young from a single mating episode. In consequence, one fertilized female has the potential to colonize a new habitat, increasing the invasive potential of these fishes still further.

Hong Kong's native fishes are all egg layers. Only one species, the ricefish (*Oryzias curvinotus*: Oryziidae) has remotely comparable breeding habits to poeciliids. In this case, the eggs are carried in a bunch at the vent of the female until they hatch or are brushed off onto aquatic vegetation. Ricefish, and other native fishes, are supplanted by poeciliids in streams which have been polluted, channelized or otherwise altered by man. While a lack of historical data precludes detailed analysis of changes in the status of indigenous fishes, circumstantial evidence suggests that interactions (competition or predation) with exotics in man-modified habitats may have caused the decline of native species. However, poeciliids are relatively unsuccessful at establishing themselves in undisturbed streams; where they do occur in such habitats it is usually at sites where the current is slight and the riparian vegetation has been cut or cleared, allowing the growth of algae and trailing grasses and herbs (e.g., *Leersia hexandra* and *Commelina nudiflora*). Poeciliids are almost invariably absent from stretches where the riparian vegetation is intact and shades the stream bed. As with plants in terrestrial environments, it is clear that alteration or degradation of stream habitats permits the proliferation of alien species.

Other alien fishes in Hong Kong freshwaters are species with some economic (aquaculture) value. Among them are nine species of carp (Cyprinidae) translocated from elsewhere in China, and, in Hong Kong, confined to reservoirs and fish ponds. African tilapias (*Oreochromis mossambicus* and *O. niloticus*: Cichlidae) have escaped the confines of aquaculture ponds and are found in disturbed streams throughout the Territory. Their spread can be attributed to the fact that these tilapias are tolerant of seawater and can swim along the coast from one stream to another. Moreover, both species are mouthbrooders and exhibit a degree of parental care. The eggs are laid and fertilized in a circular pit that is excavated by the male, and the female then incubates the eggs and newly-hatched young in her mouth. After becoming free swimming, the young stay with the mother for a short while and will retreat into her mouth when danger threatens. The success of tilapias as invasive species in Hong Kong (and elsewhere in Asia) could stem from the fact that there are very few fishes in the region with comparable breeding habits.

Considering the extent of the aquarium fish trade in Hong Kong and the fact that, until recently, many exotic fishes were cultivated for the trade in outdoor ponds, it is surprising that there are not more aliens in Hong Kong streams. While some Central American cichlid fishes can be collected from reservoirs on Hong Kong Island, the populations do not appear to be self-sustaining in the long term. However, the red-eared terrapin (*Pseudemys scripta*), which is imported by aquarium traders, seems to have colonized Plover Cove Reservoir by way of large unwanted individuals being released by their owners. Another animal with the potential to establish feral populations in the same manner is the Japanese freshwater crayfish (*Procambrus clarkii* Decapoda: Cambridae). This crustacean was introduced into Jiangsu Province in China during the 1930s, and is now cultured from Beijing to Guangdong Province. Self-sustaining, feral populations are present in many areas. These animals are imported regularly for sale as aquarium pets, and the potential for unwanted animals to be dumped in local streams is considerable.

Other freshwater invertebrates which are already established locally include a Brazilian snail, *Biomphalaria straminea* (Planorbidae) (Fig. 5G), which, in its native range, is an intermediate host of the blood fluke (*Schistosoma mansoni*) that causes bilharzia (schistosomiasis) in humans. This snail is thought to have been imported along with ornamental plants for aquaria. *Biomphalaria* was first noticed in Hong Kong by an amateur malacologist, A.J. Brandt, during late 1973 when a few snails were collected from a small stream in the lower Lam Tsuen Valley. Subsequent colonization of the Lam Tsuen River probably occurred during 1977 and, by 1980, the species had become established widely in the lowland areas of the New Territories. The spread of this snail throughout the New Territories from what appears to have been a single introduction is remarkable, and it has dispersed into adjacent areas of the People's Republic of China. This is facilitated by the fact that *Biomphalaria* abounds in wet vegetable fields (especially where water spinach [*Ipomoea aquatica*] is grown), and may be spread inadvertently on muddy boots or farming implements. These snails have been intercepted in consignments of aquarium fish exported from Hong Kong into Australia; thus the globalization of the species continues.

In addition to colonizing the New Territories rapidly, *Biomphalaria* has prospered in the new environment, becoming especially abundant in slow-flowing streams, irrigation ditches and agricultural channels, where densities can exceed 20 000 individuals/m^2. The success of *Biomphalaria* reflects a high rate of reproduction, at least during the summer months. The mean generation time is only 26 days, and individuals (which are hermaphroditic) produce 10–70 eggs/day throughout their 20-week reproductive period. *Biomphalaria* has a rather generalized diet and thrives in organically-polluted habitats where other snails cannot persist. Such attributes have allowed colonization of a range of freshwater habitats, particularly those which have been polluted or otherwise altered by human activities.

The apple snail *(Pomacea lineata:* Ampullaridae), also originates from South America, but it is not clear whether this animal was introduced directly into Hong Kong or, first, into China (where it is used as a food in southern provinces) thence spreading south across the border. *Pomacea* may also owe some of its success to a habit of consuming the eggs and newly-hatched young of potential snail competitors. Another mollusc invader, the freshwater mussel, *Limnoperna fortunei* (Mytilidae), which invaded Hong Kong by way of the pipeline bringing the supply of raw water from China, can be thought of as a species whose range was extended by man-induced habitat changes. The snail *Physella acuta* (Physidae) (Fig. 5F), also appears to be an introduction; it is thought to have originated in North America from where, with man's inadvertent assistance, it has colonized freshwater habitats in various parts of the tropics.

Hong Kong streams are also host to exotic plants, including the Brazilian water hyacinth (*Eichhornia crassipes* — first noticed by naturalist A.H. Crook in 1906), which is infamous for clogging waterways in the tropics and subtropics, and the floating fern *Salvinia molesta* (also from Brazil) which is a nuisance weed in Southeast Asia, India, Australasia and Africa. *Salvinia* was widespread in paddy fields prior to the Second World War but, with the decline and cessation of rice farming during recent decades, has become rather rare. It may have been introduced into Hong Kong as an ornamental aquarium or pond plant, and was present from at least 1912. Like water hyacinth, however, *Salvinia* is widespread in Asia, and establishment in Hong Kong due to range extension by exotic Southeast Asian populations cannot be ruled out. Water hyacinth spread may have been enhanced by its use as livestock (especially pig) feed, but *Salvinia* is unpalatable to mammalian herbivores. Clonal propagation (asexual reproduction) is likely to have facilitated the spread of both species; if plants are broken apart into separate pieces each has the potential to grow into a complete organism. Dispersal is aided by the high mobility of the plants, made possible by air-filled tissue (aerenchyma) in the stems and leaves of the plants which makes them buoyant. These weeds can therefore float with wind or water currents to unoccupied waters.

The impact of alien species

Why do exotic or alien species matter? Surely, the establishment of exotic organisms is a 'good thing' as it will increase the biological diversity of Hong Kong's hillsides, streams and rivers? The answer to these questions depends upon two things: where the exotic species establish themselves, and what they do when they get there.

Although all habitats in Hong Kong have been modified by man, definite exotics are mostly confined to those areas where human influence is strongest and most persistent. In general, this means the coastal lowlands. Indeed, in most residential and industrial areas, as well as sites used for intensive farming, exotic species dominate the biota. In contrast, recognizable exotics are rare or absent in most upland streams and hillside communities. Thus the majority of exotics are found in those places where the native flora and fauna has already suffered most as a result of human activities. The important question is therefore: are introduced species merely occupying vacant ecological niches in environments outside the range of tolerance of the native biota, or are they contributing to the decline of the natives?

There are numerous examples worldwide of introduced species bringing about the extinction of native organisms. The most dramatic have involved predators. An extreme example is the deliberate introduction of the fish-eating Nile perch (*Lates nilotica*) to Lake Victoria, in East Africa, causing the extinction of dozens of species of small endemic cichlid fishes. Introduced cats and snakes have had a similar effect on island bird faunas. Other cases have involved the introduction of grazers or competitors. It is noticeable, however, that the most dramatic problems with exotic introductions have occurred on islands or in island-like habitats, such as Lake Victoria. The impact of exotics on continental biotas (with the exception of Australia, which is also an island — albeit a large one) has been far less drastic.

In Hong Kong, as mentioned above, there is circumstantial evidence for the reduction of native fishes by introduced poeciliids, especially mosquito fish which will prey upon small fishes and harass and nip the fins of larger ones. For example, some minnows such as the endemic *Aphyocypris lini* (Fig. 18), which has not been recorded in Hong Kong since 1983, may have been eliminated through competition or predation, while ricefish can be collected only from the few habitats that poeciliids have yet to colonize. The scarcity of the endemic Romer's frog (*Philautus romeri*), which is confined to a few offshore islands, may also reflect mosquito fish predation upon the small tadpoles. Apart from the dramatic case of the pinewood nematode, there are no obvious examples of similar impacts in terrestrial habitats, but this may simply reflect the absence of local research on this question.

Where exotic species dominate a habitat, their influence on the environment for other species can be overwhelming. For example, the rapid clonal growth of exotic floating plants allows them to cover all or part of the water surface. Reduced light penetration must inhibit the development of submerged plants and restrict gaseous exchange between water and air. Water hyacinth roots do provide a habitat for a variety of aquatic invertebrates (including gastropods, chironomid larvae, dragonfly and damselfly larvae, various beetles, and heteropteran bugs), and the fauna associated with this plant in the lower Lam Tsuen River was more species-rich than that of the bottom sediments (93 versus 19 taxa). However, it could be argued that this floating root habitat can be provided by native species such as the water

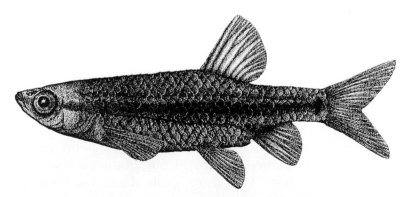

Fig. 18 Hong Kong's rarest freshwater fish: *Aphyocypris lini* (Cyprinidae); body length 27 mm. Drawing by Chong Dee-hwa.

lettuce, *Pistia stratiotes* (Araceae), or replaced by a combination of submerged and emergent plants. Moreover, the species-poor benthic fauna may have been a result of relatively low oxygen concentrations beneath the Lam Tsuen River water hyacinth beds.

In terrestrial habitats, the aggressive South American climber, *Mikania micrantha*, and several species of African grasses, similarly dominate considerable areas. Again, however, these species are occupying a habitat already highly modified by man. Interestingly, both this climber and the exotic *Lantana camara* are good sources of floral nectar for butterflies, and *Lantana* flowers throughout the year. It is thus possible that these exotics facilitate recolonization of disturbed areas by native butterflies, through provision of food for the adults. However, without a suitable larval food plant, establishment of self-sustaining butterfly populations is impossible. *Lantana* also provides a continuous supply of fruits through all but the driest months of the year and these seem to be an important source of food for native birds in disturbed areas.

At present, the impact of the numerous exotic plant and animal species established in Hong Kong is, in most cases, hard to distinguish from the direct impact of human activities on the habitats they occupy. With the obvious exception of the pinewood nematode, exotics seem mostly to be a symptom, rather than a cause, of environmental degradation. Does this mean that we should not be concerned about future introductions? We believe that we should be very concerned, for three main reasons. Firstly, we may well be underestimating the impact of the species already naturalized in Hong Kong. Ecological interactions, such as competition and predation, are notoriously difficult to study under field conditions and such studies have not even been attempted locally. Secondly, even if few already-established exotics are causing problems, we cannot assume that the same will apply to all future introductions. A single species could transform our hillsides or streams in totally unpredictable ways. Once more, the pinewood nematode can be cited as an exotic species having far-reaching consequences. Finally, many of the exotic species established in Hong Kong have the potential for a major economic impact on the more agricultural economy of South China. Pallas's squirrel will follow the pinewood nematode across the border, where it has the potential to become a pest in plantations. Many of our exotic plants are already economic weeds in China. Even if we ignore the risk to the ecology of Hong Kong, we have an obligation to ensure that the Territory does not become a stepping stone for invasion elsewhere.

9

Conservation

Despite the massive human impact described in the previous chapters — deforestation, erosion, fire, hunting, trapping, pollution and the introduction of exotic species — Hong Kong's flora and fauna are still surprisingly diverse. The Territory supports more native plant species than Britain, more mammals, more reptiles, more amphibians, more fish, more butterflies, more moths, more ants, more dragonflies and damselflies, more . . . the list could go on and on. Furthermore, terrestrial habitats in Hong Kong are, on the whole, better preserved than those in adjacent areas of mainland China. Conservation in Hong Kong is still possible and still worthwhile! The potential for conservation is highlighted by the apparent rise in species richness of our forest-adapted bird fauna has over the past few decades which has accompanied an increase in the area and maturity of plantations and secondary woodlands (see Chapter 5). Similarly, the once rare map butterfly (*Cyrestis thyodamas*: Nymphalidae) has increased in abundance and range over the New Territories during the past few years.

This is not to say that there have been no extinctions. Documented extinctions are few, but examples from this century include the tiger, the leopard, the South China red fox, the South China badger, the large Indian civet (not seen in the wild for over a decade), the ring-necked pheasant (*Phasianus colchicus*), the water monitor, a number of freshwater fishes, a whole family of stream-dwelling caddisflies (the Dipseudopsidae), and probably the floating frog

(*Ooeidozyga lima*). This list is short because massive human impacts which occurred in Hong Kong before the start of reliable records in the nineteenth century had already eliminated the most sensitive species. The majority of botanical extinctions probably occurred in the coastal lowlands, to which the most cold-sensitive species would have been confined by the more frequent frosts in the hills. Complete deforestation at low altitudes must have resulted in the loss of a substantial fraction of Hong Kong's native flora. For forest-dependent vertebrates the situation would have been even worse. Considerable botanical diversity can survive in forest fragments protected from fire and cutting but, even without hunting pressure, most vertebrates need larger areas to support a viable population.

Our existing flora and fauna, therefore, consists of species which have survived massive human impact in the past. Why are they threatened now? The basic answer is straightforward. The last few decades has seen a decline in human impact over much of the Territory as a result of rural depopulation, contrasting with a massive increase in impact on the coastal flatlands. However, this pattern of increasing concentration of population is now undergoing a partial reversal. The population of Hong Kong Island is holding steady, that of the Kowloon Peninsula is declining, and that of the New Territories rising rapidly. The massive developments on Lantau Island associated with the new airport and port facilities will accelerate this trend. It is this redistribution of people and industry to previously rural areas, coupled with rising living standards and increased mobility, that is the basic threat to wild nature in Hong Kong.

The situation is exemplified by the status of the local freshwater fish fauna, with 11 species threatened with extinction or already extinct. The endangered species seem to be confined to lowland rivers and streams — habitats which are impacted heavily by human activities. One of the threatened species, the minnow *Aphyocypris lini* (Fig. 18), is found nowhere else on earth but Hong Kong. It has not been seen in the wild for ten years and may now be extinct. In 1993, streams at Tai Ho and Tung Chung on Lantau Island supported an unusually high fish diversity, yielding 47 and 23 species, respectively (with a combined total of 53), including some threatened fishes. Unfortunately, these are precisely the streams which will be most affected by the airport development.

Britain and China were among the 153 countries which signed the Convention on Biological Diversity at the United Nations Conference on Environment and Development in June 1992 in Rio de Janeiro —

the 'Earth Summit'. This Convention came into force on 29 December 1993 after it had been ratified by 30 countries. Under the Convention, each country is required to develop strategies for the conservation and sustainable use of biodiversity. Since both Britain and China are signatories, Hong Kong is obliged to comply with the Convention both before and after 1 July 1997. Although the Biodiversity Convention requires nothing that Hong Kong should not be doing anyway, it does serve to place our local concerns in a global context.

Protected areas

Legal protection of biological diversity can be aimed either at areas (such as Mai Po Marshes) or individual species (such as the leopard cat). The major conservation advantage of targeting areas rather than species is that protected areas will include numerous inconspicuous, unknown and 'non-charismatic' species, as well as the birds, mammals, flowers and butterflies that most of us think of as 'nature'. Protected areas can also provide for other needs, such as water catchments, education and recreation. The major disadvantage is that protected areas are usually chosen by default — the areas not immediately wanted for anything else — and may thus omit many important habitats and species.

The minimum requirement for a protected area is that it should remain protected in perpetuity. The largely negative criteria used to select many protected areas also make them extremely vulnerable to encroachment by development as land values change, because no positive case need have been made for protection. This makes revoking protected status much easier. Hong Kong's current protected areas illustrate many of these points.

The Hong Kong government was very late in developing a protected area system, having previously relied on a variety of laws to control human activities on Crown Land. In the 1960s, however, it became apparent that the rapidly increasing pressures on the countryside and the urgent need for recreational outlets for the urban population made the establishment of a formal park system essential. As a result, the present Country Parks system was initiated in the early 1970s and, making up for lost time, largely completed by the end of 1979.

Today, Hong Kong's 21 Country Parks, plus the three Special Areas which are not inside Country Parks, together cover more than

40% of the Territory — an impressive proportion in view of Hong Kong's extremely high population density. Within these areas, all organisms and their habitats are legally protected. Country Park boundaries are shown in the map of Hong Kong and Figure 19. Special Areas within Country Parks receive no additional legal protection but the extra status does serve to highlight areas of particular conservation significance. Heavy recreational use of Country Parks is less of a problem than might be expected because it is concentrated in those areas provided with special facilities, such as barbecue pits, and with easy access by public transport. Moreover, the impact is confined largely to weekends and public holidays. From a conservation viewpoint, however, the Country Parks system is deficient in at least two aspects.

Firstly, there is the absence of a clear, long-term conservation management policy for Country Parks. Apart from the imaginative and successful control of recreational impact, conservation management is limited to fire prevention and the planting of trees, which are still

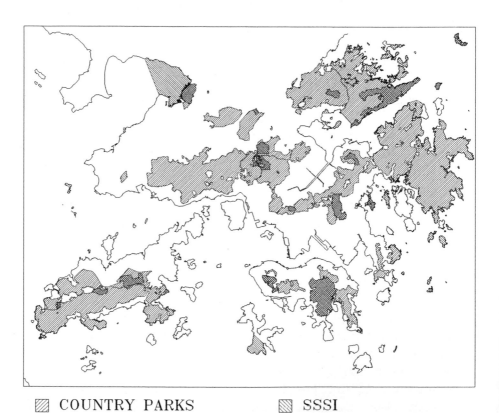

▨ COUNTRY PARKS   ▨ SSSI

Fig. 19   Hong Kong Country Parks and larger Sites of Special Scientific Interest (SSSIs).

mostly exotics. Despite considerable efforts in public education, fire prevention and fire-fighting, fire is still the major threat to the diversity and beauty of Country Parks. There is no single, simple answer to this problem but it is clear that more resources must be made available for construction of fire breaks, law enforcement, and fire control. A massive increase in the penalties for starting fires in the countryside would serve an educational function, even though enforcement is extremely difficult. The general public does not see hill fires as the disaster they really are, precisely because they are so common. Our level of tolerance must be reduced.

Tree-planting in Country Parks has been a success, in terms of numbers placed in the ground, although many trees are lost annually to fires. But the purpose of planting trees is far from clear, so the policy cannot be judged against any particular objective. From a conservation viewpoint, pure stands of exotic species in neat rows are the least valuable form of tree cover. Mixing two species of trees, to provide greater structural complexity, is an improvement, but the greater use of native mixtures, which provide food and shelter for indigenous wildlife, should be the aim (see also Chapter 5: Plantations). Even with native mixtures, care must be taken to choose species with the desired ecological characteristics: not simply those species which are easiest to produce in large numbers.

Except where rapid establishment of tree cover is needed to prevent erosion in catchment areas (and trees are not always the best ground cover for this purpose), or to shade recreational facilities, woodland quality should carry at least as much weight as quantity when decisions about planting are made. In view of the high cost of planting trees, it makes sense to choose the sites most suitable for tree growth, rather than the trees most suitable for poor sites, as often happens at present. A given area of trees planted so as to extend an existing woodland is more valuable to wildlife than an isolated 'island' of the same area planted in an expanse of grassland. Furthermore, encouraging natural succession by fire prevention may often be more effective than the same amount of money spent on tree planting.

Conservation management could be taken a step forward if the annual reports of the Conservation and Country Parks Branch of the Agriculture and Fisheries Department were framed in terms of specific conservation objectives, rather than simply listing numbers of visitors, tons of litter collected, fires fought and trees planted. However, little can be achieved without a very large increase in the resources available to the Country Parks administrators. Current funding and staffing levels seem to be based on the assumption that managing Country Parks is essentially a passive activity, and an unskilled one at that. Conservation funding in Hong Kong has clearly suffered from a widespread belief that 'The Environment' means air and water quality — the responsibility of the well-staffed and well-funded Environmental Protection Department — and that biological diversity should be and is looked after by voluntary organisations, such as the World Wide Fund for Nature (Hong Kong). The fact that Country Parks and Conservation are buried within a department responsible for agriculture and fisheries — two areas of declining economic importance — can do nothing to help this image problem.

The management of Country Parks for conservation is also in potential conflict with the Country Parks Board's legal obligation under the ordinance to encourage recreational use of the parks. This has been interpreted by some members of the Board as justification for the construction of golf courses and other ecologically-damaging facilities within Country Parks, although it is clear this was not the intention of those who established the system two decades ago.

The second deficiency of the Country Parks system is in its coverage of important habitat types. The present Country Parks consist of areas of unoccupied Crown Land for which no significant alternative use existed at the time they were established or was envisaged for the future. Many of the areas chosen were water catchments for reservoirs which had previously been protected under the Waterworks Ordinance and were the main targets for post-war afforestation efforts. The aims of the planners were primarily to preserve Hong Kong's countryside for human use rather than conservation of biological diversity. Country Parks had to be seen as serving the needs of the general public rather than simply a small minority of naturalists. Moreover, the boundaries had to be acceptable to villagers who feared a loss of traditional rights.

What this has meant in practice is that not only were village lands, including abandoned paddy fields and *feng shui* woods, usually excluded, but Country Park boundaries often omitted the surrounding valley bottoms and lower hill-slopes as well. Freshwater wetlands and *feng shui* woods are the most obvious omissions from the present Country Parks system, but lowland habitats in general are very poorly represented. These omissions are not always obvious when walking in the countryside because so many rural villages have been abandoned, but these areas are protected by isolation rather than law and are thus always vulnerable to future development. The recently proposed development of the Sha Lo Tung basin exemplifies this point, but near-wilderness areas, such as Tai Long Wan in the eastern New Territories, are also at risk. In some places, relatively minor adjustments to the existing boundaries could ensure protection of valuable habitats.

The problems of covering all important habitats adequately are heightened by the fact that biological diversity in Hong Kong tends to be concentrated in a few small areas. Unfortunately, we cannot always predict where these areas will be. Moreover, high diversity of one group of organisms at a particular site is not generally correlated with high diversity in the other groups. For example, almost half of the species of stream fishes recorded from Hong Kong (including some known from nowhere else in the Territory) inhabit Tai Ho Stream on Lantau, but the site is not unusually rich in freshwater invertebrates or aquatic plants. Similarly, there are around 90 species of dragonflies and damselflies known locally, more than half of which have been recorded within the Sha Lo Tung Basin — the most species-rich site in the Territory for Odonata. In this case, at least, insect diversity correlates with the presence of two of the three amphibians which are protected by law in Hong Kong. However, at Tai Tong (near Yuen Long), an impressive total of 36 dragonflies and damselflies (including one new to science) has been collected from a short stretch of stream draining a valley which, in all other respects, is quite unprepossessing. Chek Lap Kok, site of Hong Kong's new airport, provided habitat for a wide array of reptiles and amphibians — including the endemic Romer's frog — but the terrestrial biota was not outstanding in other respects. Elsewhere in the Territory, it has become apparent that data from botanical studies do not allow prediction of the best sites for butterflies, nor does bird diversity correlate directly with tree species richness. The patchwork of nature sets us conservation challenges which will not be met easily. It is nevertheless clear that we need to know much more about what is where so that we can set priorities for conservation.

Some, but by no means all, of the gaps in the Country Parks system are filled by other kinds of protected area. Outside the scope of this book is Mai Po Marshes, Hong Kong's premier reserve, protected as a restricted area under the Wild Animals Protection Ordinance. In addition, the recently-amended Town Planning Ordinance provides for the designation of '. . . coastal protection areas, Sites of Special Scientific Interest (SSSIs), green belts or other specified uses that promote conservation or protection of the environment.'

Of these categories, only SSSIs were in existence before 1991. The 49 current SSSIs were declared (from 1975 onwards) when such status was purely an administrative matter and had no legal standing. They are a very mixed lot in terms of both size and quality, and many simply highlight areas which are also protected in other ways. SSSI status is probably the best way of providing protection to small areas of special conservation value outside the Country Parks system, particularly if they are on private land. We suggest that all the best *feng shui* woods and freshwater wetlands should be SSSIs, as well as representative samples of other lowland habitats. At the same time, the current list should be 'weeded' of sites which do not deserve protection.

Unfortunately, the Town Planning Ordinance does little more than ensure that the existence of SSSIs is known to planners. Eventually, SSSIs and the new conservation and landscape protection categories will be shown on Outline Zoning Plans, accompanied by an indication of what activities can and cannot be carried out in them. However, these plans will be relatively easy to change while enforcement procedures against unauthorised developments are both weak and complex. Moreover, the majority of the Territory will not be covered by such plans in the immediate future. This is especially important in view of the fact that isolated, lowland areas (such as parts of Tai Long Wan) lie outside Country Park boundaries, and are susceptible to the ambitions of developers. We hope to be proved wrong but, at present, we see no evidence that the Town Planning Ordinance will have a significant positive impact on conservation.

Additional legislation is urgently required to clarify the status of SSSIs and the other conservation categories mentioned in the Town Planning Ordinance. Particularly valuable would be a legal designation for areas which deserve protection from human impact but have neither the special conservation value needed to justify SSSI status (which we believe should not be diluted) nor the requirement for active management implied in Country Park status. The fact that the declaration of such conservation areas would have few immediate financial or manpower implications would greatly facilitate the protection of new areas.

In addition to the specific areas mentioned in existing legislation, the Forests and Countryside Ordinance prohibits the felling, cutting or burning of trees and other plants within all forests on Crown Land. This Ordinance is the most recent of a long series of attempts to protect forests in Hong Kong, stretching back to the beginning of large-scale afforestation in the 1870s. Originally called the Forestry Ordinance, the title was changed in 1974 and '. . . the protection of the countryside' added to the previous aims. It is not clear what additional protection this law gives to upland forests, as most of these are within Country Parks, but it gives significant extra protection to mangrove areas.

Protected species

As well as protected areas, where all species are — at least in theory — safe, Hong Kong also has protected species which cannot be killed, injured or collected anywhere in the Territory. A protected-species list is a useful supplement to a protected-area system, not only for the conservation of species which occur partly or exclusively in unprotected areas, but also as an aid to enforcement. Someone found with a recently-killed barking deer, for instance, cannot escape prosecution by claiming it was caught outside Country Park boundaries.

There have been laws to protect wild birds in Hong Kong since the Preservation of Birds Ordinance of 1870 but, until recently, it was permissable to kill both 'game birds', such as wild ducks, snipe and pigeons, and 'vermin', which included magpies and birds of prey. Indeed, in 1909, the Government Gazette included a notice to the effect that '. . . with a view to preserving song birds in this Colony, His Excellency the Governor will be glad if all holders of game licenses . . . will destroy magpies whenever the opportunity arises.' No mammal species was protected until the Wild Animals Protection Ordinance of 1936, which initially listed only the pangolin and the otter.

Today, the Wild Animals Protection Ordinance covers all wild birds, all wild mammals except rats, mice and shrews (but including the introduced Pallas's squirrel), selected reptiles (including the locally-extinct water monitor), three amphibians, and a single insect — the birdwing butterfly (Table 5). The list probably includes most of the species that could be threatened by trapping or collecting. Properly enforced, it would go a long way towards eliminating this particular threat. However, listing under this ordinance provides no direct protection against the more insidious threats of habitat destruction and, for aquatic species, pollution.

Legal protection of named plant species started in 1913 with a regulation made under the Licensing Ordinance which prohibited the hawking or 'possession with a view to hawk' of nine taxa: all rhododendrons, the lady's slipper orchid (*Paphiopedilum purpureum*), the Chinese New Year flower (*Enkianthus quinqueflorus*), *Ixora*

Table 5   Animal species protected under the Wild Animals Protection Ordinance.

Scientific name	Comments
Mammals	
Chiroptera	All bats; 21 species have been recorded locally.
Primates	All primates; one native and three exotic species of macaques (plus hybrids) occur in Hong Kong.
Manis pentadactyla	Chinese pangolin.
Hystrix brachyura	Chinese porcupine.
Sciuridae	All squirrels; only the exotic Pallas' squirrel (*Callosciurus erythraeus*) occurs in Hong Kong today.
Cetacea	All species of whales, dolphins and porpoises.
Vulpes vulpes	South China red fox; extinct in Hong Kong.
Herpestes spp.	All mongooses; the crab-eating mongoose (*H. urva*) and the Javan mongoose (*H. javanicus*) have been recorded locally.
Paguma larvata	Masked palm civet.
Viverricula indica	Small Indian civet.
Viverra zibetha	Large Indian civet; almost certainly extinct in Hong Kong.
Lutra lutra	Otter.
Melogale moschata	Chinese ferret badger.
Felis bengalensis	Leopard cat.
Dugongidae	All dugongs. One species, *Dugong dugon*, must have been present in the Hong Kong region but is now certainly extinct.
Muntiacus reevesi	Barking deer.
Birds	
Aves	All birds; 421 species have been recorded locally.
Reptiles	
Testudines	All turtles and terrapins; 11 species have been recorded locally.
Python molurus bivittatus	Burmese python.
Varanus salvator	Water monitor; almost certainly extinct in Hong Kong.
Amphibians	
Paramesotriton hongkongensis	Hong Kong newt; apparently endemic to Hong Kong.
Amolops hongkongensis	Hong Kong cascade frog; aparently endemic to Hong Kong.
Philautus romeri	Romer's tree frog; apparently endemic to Hong Kong.
Insects	
Troides helena	Birdwing butterfly.

chinensis, the Chinese lily (*Lilium brownii*), the fragrant litsea (*Litsea cubeba*), the Hong Kong orchid tree (*Bauhinia blakeana*), *Manglietia fordiana* and *Pavetta hongkongensis*. Under the Plants Ordinance of 1920, this list was gradually extended until, by 1936, it covered more than a quarter of the total flora, including all ferns, orchids, and hollies (*Ilex* spp.), all Fagaceae, and such widespread species as *Dianella ensifolia*, *Rhaphiolepis indica*, *Sterculia lanceolata* and *Smilax china*.

Plant species currently receive protection under the Forestry Regulations of the Forests and Countryside Ordinance. Although it has been extended recently, the list of protected species is much shorter than it was in 1936 (Table 6). Listed plants include all native orchids, camellias, rhododendrons, tree-ferns and a variety of other species. This list is far less effective than the vertebrate portion of the Wild Animals Protection Ordinance because it covers a smaller proportion of the Hong Kong species and includes largely plants of high-altitude areas which are almost all in Country Parks. Thus the protected plant list cannot be used as a way of identifying unprotected areas which should not be developed. The major criterion for inclusion on the list is the potential threat from collectors, rather than rarity or endangerment from other causes. This betrays the origin of Hong Kong's plant protection laws in attempts to control sales by hawkers.

It is clear that Hong Kong's protected species legislation is out of step with modern Hong Kong. It is illegal to trap a civet or uproot an orchid but acceptable to pave, concrete over or otherwise degrade their habitat. A more useful approach might be an official list of endangered species and a legal requirement that the welfare of these species is considered in all developments. The presence of Romer's frog did not stop the Chek Lap Kok airport, nor would we have expected it to. However, we do believe that the presence of endangered species should be given weight when there is a choice of sites or the development is non-essential. At present, this is not required.

Environmental impact assessment

In addition to protecting areas and species, there is a third way in which undesirable impacts on Hong Kong's natural landscape can be reduced. All developments, even on private land, require government permission before they can go ahead. If the project is likely to have an adverse impact on the environment, the government can require that

Table 6 Plant species (excluding orchids) protected under the Forestry Regulations of the Forests and Countryside Ordinance. In addition, all species of wild orchids are protected.

Scientific name	English common name
Ailanthus fordii	Ailanthus
Amentotaxus argotaenia	Amentotaxus
Angiopteris evecta	Mule's foot fern
Aristolochia tagala	Indian birthwort
Asplenium nidus	Bird's nest fern
Camellia assimilis	
C. caudata	
C. crapnelliana	Crapnell's camellia
C. granthamiana	Grantham's camellia
C. hongkongensis	Hong Kong camellia
C. kissii	
C. oleifera	Oil tea
C. salicifolia	Willow-leaved camellia
C. sinensis var. waldenae	Walden's camellia
Cyathea podophylla	Tree-fern
C. hancockii	Tree-fern
C. lepifera	Tree-fern
C. spinulosa	Spiny tree-fern
Dendrobenthamia hongkongensis	Hong Kong dogwood
Drosera peltata	Crescent-leaved sundew
Enkianthus quinqueflorus	Chinese New Year flower
Illicium angustisepalum	Lantau star anise
I. dunnianum	Dunn's star anise
I. leiophyllum	Wild star anise
I. micranthum	Small-flowered star anise
Illigera platyandra	Illigera
Impatiens hongkongensis	Hong Kong balsam
Iris speculatrix	Hong Kong iris
Keteleeria fortunei	Keteleeria
Lagerstroemia fordii	Ford's crape myrtle
L. subcostata	South crape myrtle
Lilium brownii	Chinese lily
Magnolia championi	Hong Kong magnolia
Magnolia fistulosa	
Manglietia fordiana	Manglietia
Michelia maudiae	Maud's michelia
Nepenthes mirabilis	Pitcher plant
Pavetta hongkongensis	Hong Kong pavetta
Platycodon grandiflorum	Balloon flower
Rehderodendron kwangtungense	Guangdong rehdertree
Rhododendron championae	champion's rhododendron
R. farrerae	Farrer's azalea
R. hongkongense	Hong Kong azalea
R. simiarum	South China rhododendron
R. simsii	Red azalea
R. westlandii	Westland's rhododendron
Rhodoleia championi	Rhodoleia
Tutcheria spectabilis	Tutcheria
Schoepfia chinensis	Chinese schoepfia

an Environmental Impact Assessment (EIA) is done before work starts. The EIA identifies potential environmental problems and suggests ways in which these problems can be reduced. If environmental impacts cannot be reduced to an acceptable level, the project may have to be abandoned.

EIAs in Hong Kong have usually concentrated on pollution and the visual impact but impacts on the flora and fauna are increasingly being included. In theory, the EIA process could be a powerful weapon in the defense of nature in Hong Kong. In practice, the benefits have been less than they could be. To start with, there is, as yet, no legislation which deals specifically with the EIA process, making the enforcement of correct procedures very difficult. Secondly, the time allowed for completion of the EIA is nearly always far too short, so important impacts may easily be missed. Finally, our ignorance about so much of Hong Kong's ecology, and the narrow dissemination of what little we do know, makes the task of predicting impacts and assigning importance to species and habitats extremely difficult. In such a situation, the 'burden of proof' should be placed on the developer to show that the proposed project will not damage the environment. In reality, however, the onus is on opponents of the project to prove that it is environmentally unacceptable.

Even when an EIA has been undertaken, and a series of recommendations made as to how ecological impacts can be avoided, reduced or mitigated, there is little attempt to ensure that contractors or managers comply with these recommendations. Thus construction waste is dumped in sensitive areas, bulldozers plough up vegetation and push earth into streams, and residents of new developments cut down trees and modify the surrounding countryside with apparent impunity.

One way of dealing with the problem of lack of compliance with EIA recommendations would be to require that developers or managers deposit a significant sum of money, in the form of a bond, before work on, or occupation of, a particular project is permitted. Should the developer or manager fail to comply with the conditions laid down in EIAs, or allow the surroundings to deteriorate to an unacceptable extent, then some or all of the bond money could be forfeited and used to mitigate the environmental damage and restore impacted habitats. Another possibility would be a legal requirement for project developers to undertake off-site mitigation in cases where their work leads to the destruction of important habitats. For example, damage to native woodland should be mitigated by compensatory planting (at a ratio of, say, three hectares planted to every hectare destroyed) to extend existing woodland in other parts of the Territory. The same principle could be applied to wetlands: drainage of one wetland could be mitigated by purchasing and setting aside another area, thereby protecting the latter site from development. Any necessary habitat maintenance could be funded by the interest from a sum put in trust by the developer.

Restoration

The present state of Hong Kong's countryside shows the effects of centuries — perhaps millennia — of human impact. Rich as they are, all terrestrial habitats in modern Hong Kong must be a pale reflection of the exuberant diversity of the primeval landscape. Can this diversity be restored?

Even if it were feasible, restoration of the entire pre-human fauna, complete with tigers, leopards, elephants and crocodiles, would obviously not be compatible with Hong Kong's present human population density. The problems with the expanding macaque populations in the Kowloon Hills show that even a medium-sized herbivore can conflict with other uses of the countryside. However, these incompatible species are exceptions, and, in their absence, their ecological roles could probably be substituted by careful human intervention. A primeval forest with 99% of its original biota would certainly be an attractive target. Unfortunately, as discussed in previous chapters, we have very little information on what the primeval forest and its fauna was like.

We suggest that a practical long-term target for those parts of the Territory which escape urbanization would be a landscape supporting the maximum possible diversity of native species. Such a landscape would probably resemble the primeval landscape more than modern Hong Kong, but it would also differ in important ways. Centuries of soil erosion have irreversibly changed the physical environment on Hong Kong's hillsides and it is doubtful if closed forest could ever be re-established on the worst-affected sites. In any case, some of our most diverse communities, including shrublands and most freshwater wetlands, are the results of past human impact yet still worth preserving.

Restoration of streams will be especially difficult because much of the flow to the lower reaches is diverted into tunnels and storage reservoirs. Where this extraction is not undertaken by the Water

Supplies Department, villagers have impounded streams to collect water for domestic use and irrigation. This water transfer affects the ecology of lowland streams and rivers profoundly, and the deleterious effects are magnified many times by pollution. It is in these perturbed rivers and streams that exotic species have established themselves and flourished (see Chapter 8). The interaction of these factors is likely to be more damaging than the simple sum of their effects, and may rule out restoration of streams to their original state. Although the government is taking steps to reduce the input of livestock wastes into rivers, the rapid population growth and development of the New Territories do not augur well for the future of Hong Kong's running waters. Moreover, it will never be possible to halt extraction of water from streams, nor to exterminate all exotic species. The best that we can hope for is a return to relatively clean freshwaters, albeit with highly-modified communities of animals and plants.

One controversial question which cannot be avoided is that of introducing species from outside Hong Kong. Few would argue with the reintroduction of species whose past presence in Hong Kong is well-documented, such as the South China red fox (after quarantine for rabies), the water monitor and ring-necked pheasant. However, as noted above, few species are in this category: most extinctions of the lowland forest biota, in particular, must have taken place centuries before any records were kept. The vertebrate fauna of Hong Kong's existing forests is clearly impoverished in comparison with other Asian forests at similar latitudes. There are no native squirrels, probably none of the original primate populations, no bears, no pheasants or jungle fowl, no big cats, no large deer. . . . The almost complete deforestation of the South China lowlands and continued pressure from hunting and trapping makes it extremely difficult for recolonisation to take place spontaneously. Should we (re)introduce the presumed missing taxa? And what of species which are extinct throughout their original range, not just in Hong Kong? Would introduction of a related species as an ecological substitute be acceptable?

These are not questions which have to be answered immediately. We can go a long way along the path of ecological restoration with the species already present in Hong Kong, plus the careful reintroduction of recent, known losses. Ultimately, however, the introduction of species from adjacent parts of China to fill obvious gaps in our biota, and the replacement of regionally extinct species with ecological equivalents from further afield, will have to be considered. Perhaps we will never see elephants in Sai Kung Country Park or long-nosed crocodiles in the Lam Tsuen River, but flying squirrels, sambar deer and, perhaps, even gibbons or langurs at Tai Po Kau are real possibilities.

What we see around us, on the hills and in the streams, are the scattered pieces of Hong Kong's original biota. Not all the pieces, but enough to make saving them our top priority. By themselves, however, these pieces are not enough to restore the original whole. Whether we like it or not, we are the stewards of Hong Kong's physical and biological environment — nature alone cannot heal the wounds that man has inflicted over the centuries. Hong Kong's history can be seen as a vast ecological experiment, the results of which have formed the major subject of this book. But the time for passive observation is past. Saving what we have and restoring what we have lost will require more than knowledge: it will also require action.

Glossary

Biological terms which have not been defined in the text, or which appear in several different places, are defined below. The definitions are operational and refer to the manner or context in which terms have been used in this book.

abaptation The process by which the characteristics of an organism were determined by the action of natural selection on its ancestors. *Cf.* adaptation.

abiotic Non-living; usually applied to physical and chemical factors in an environment. *Cf.* biotic.

adaptation A genetically-determined characteristic that enhances the ability of an organism to survive and reproduce in its environment; usually applied to beneficial features which are believed to have evolved as a result of natural selection acting on previous generations. The prefix *ad-* (Latin 'to') gives the erroneous impression that an organism is matched to present day environments. In reality, the characteristics of organisms are the result of the influence of past environments on the survival and reproduction of their ancestors. *Cf.* abaptation.

alien species See exotic species.

allochthonous Material entering one habitat (such as a stream) but derived from another (the adjacent land); i.e., exogenous. *Cf.* autochthonous.

alluvium An inorganic deposit of sand, silt and gravel which has been transported to the site by water. *Cf.* colluvium.

altruistic Refers to an animal which undertakes actions which benefit other individuals at some cost to itself; i.e., the altruist is usually seen as making a sacrifice, such as foregoing reproduction while helping another individual to rear its young.

amphibiotic Animals which spend part of their life cycle in water and the rest on land; includes amphibians and many stream insects which have aquatic juveniles and terrestrial adults.

annual plants Species which can germinate, grow and set seed within one year; living for one year or less.

aposemete A poisonous, dangerous or distasteful animal which is brightly coloured to advertise its unpleasant attributes; species with warning colouration are said to exhibit aposematism.

arboreal Living in trees.

aquatic Living in water, as opposed to terrestrial.

autochthonous Material produced within a habitat (such as organic material produced as a result of photosynthesis by algae growing in a stream); i.e., endogenous. *Cf.* allochthonous.

autotrophic Describes organisms which synthesize organic matter from inorganic materials, typically by the process of photosynthesis; also used of communities which are net producers of organic material and do not depend on allochthonous inputs. *Cf.* heterotrophic.

Batesian mimicry Resemblance of a palatable or harmless species (the mimic) to an unpalatable or harmful species (the model, usually an aposemete); it is assumed that this deceives predators so conferring protection on the mimic. *Cf.* Müllerian mimicry.

benthos The organisms which live on or in the bed of rivers, streams and reservoirs; bottom-dwelling or benthic organisms.

biomass The weight of living material (but including dead tissues of the bark and heart wood of living trees) usually expressed as dry weight per unit area (as are all biomass figures in this book).

biota All of the living organisms occurring within a particular area; e.g., Hong Kong's biota.

biotic Living; pertaining to the biological factors in an organism's environment, such as its natural enemies, breeding partners, competitors, prey and so on. *Cf.* abiotic.

boundary layer The thin, relatively slow-moving layer of water just above the bed of a river, where speed of flow is reduced by friction.

Carboniferous The geological period from 270 to 220 million years ago.

canopy The uppermost continuous stratum or layer of vegetation in a forest. In effect, the crowns of the tallest trees. *Cf.* understorey.

carnivore An animal which eats the flesh of other animals.

carrion The dead bodies of animals.

cellulose A complex polymer of glucose; i.e., a polysaccharide; the main constituent of plant cell walls.

channelization Engineering work involving straightening the course of a river or stream, whereby the complex natural pattern of meandering or braided flow and irregular banks is replaced by a straight channel with uniform, smooth banks.

chitin A complex polysaccharide (glucose polymer) which is laid down in chains and forms the major component of insect cuticles (exoskeletons).

climax A community which has reached an apparently unchanging, steady state under particular environmental conditions; the presumed end point of a successional sequence.

coevolution Reciprocal evolution of two interacting species whereby evolutionary change in one leads to a change in the other.

collector-filterer   An aquatic animal which filters fine particles of organic matter suspended in the water flowing past it; also known as a filtering-collector or filter-feeder.

collector-gatherer   An aquatic animal which gathers fine particles of organic matter from the surface of bottom sediments. *Cf.* deposit-feeder.

collectors   Combined term for collector-gatherers plus collector-filterers.

colluvium   An inorganic deposit, typically of soil, cobbles and boulders, which has been transported to the site by gravity; e.g., by a landslide. *Cf.* alluvium.

community   The organisms in a particular place or habitat; the species occurring together in the same place at the same time; the biotic component of an ecosystem.

community respiration   The respiration rate of an entire community; usually expressed in terms of oxygen consumed per unit area per unit time.

community structure   The species present in a community and their relative abundances; also used to refer to the composition of a community in terms of the functional roles or feeding methods of the constituent species. See functional classification and functional feeding groups.

competition   The struggle among organisms for a larger share of a limiting resource; intraspecific competition occurs among members of the same species, and interspecific competition between members of different species.

conglomerate   Cement-like rock made up of rounded fragments varying greatly in size.

constraint   Any restriction on the response of a population to selection.

consumer   An organism which feeds on other organisms or dead organic matter. See heterotrophic.

coprophagous   Eating faeces.

corolla   Collective name for the petals of a flower.

Cretaceous   The geological period from 140 to 65 million years ago.

crypsis   Concealment of an animal by blending with its surroundings; cryptic animals are usually coloured to resemble their food plant or resting place, and so escape detection by predators.

cultivar   A variety of plant produced by human selection in cultivation.

cytoplasm   The living contents of plant and animal cells.

decomposition   Degradation of complex organic matter into simple inorganic and organic compounds.

deposit-feeder   An aquatic animal which consumes fine particles of organic matter that are on and within the bottom sediments. *Cf.* collector-gatherer.

detritus   Dead organic matter; the particulate remains of plants (usually) and animals.

detritivore   An animal which eats detritus and the associated microorganisms.

Devonian   The geological period from 413 to 365 million years ago.

diapause   A resting period in the life cycle of many insects, characterized by reduced metabolism and suspended growth.

direct development   Applied to those freshwater crustaceans where the planktonic larval stages are suppressed and larval development takes place inside a large-yolked egg — as a result, hatchlings resemble a miniature adult in form and habit.

diversity   A measure of the variety of species in a community, taking into account both the number of species present and their relative abundances; one community is more diverse than another if it has more species and/or the relative abundances of the species are more nearly equal. See species richness.

drainage basin Land that drains into a particular river, stream or reservoir.

drift Benthic animals carried downstream by the current; **behavioural drift** arises as a direct result of behaviour and may be actively initiated by stream animals so serving as a means of dispersal within the habitat; **passive drift** occurs when animals are accidentally dislodged and swept away by the current during normal conditions of flow; **catastrophic drift** refers to the washout of animals during flood flows or spates when the stream bed is physically disturbed by the current.

ecology The scientific study of the interactions that determine the distribution and abundance of organisms; also defined as the study of relationships (or interactions) between organisms and their environment, or as the study of adaptations, or as scientific natural history.

ecosystem An interacting ecological unit comprising the biotic community and abiotic environment within a defined area; ecosystem boundaries are usually arbitrary and the entire earth may be thought of as a single ecosystem, the biosphere.

EIA See environmental impact assessment.

emergence Transformation to the adult stage in insects; in stream insects, this involves leaving the water and (usually) a nuptial flight.

endemic Confined to a certain region; e.g., Romer's frog (*Philautus romeri*) is endemic to Hong Kong.

environment The surroundings of an organism, including the biotic and abiotic components.

environmental impact assessment (EIA) An estimate, appraisal and evaluation of the total effect of a man-made environmental change, containing recommendations as to how potential impacts might be minimized or mitigated by compensatory activities.

eukaryote An organism whose cells have a distinct nucleus surrounded by a membrane, with discrete organelles in the cytoplasm, and DNA as the genetic material (e.g., protozoans, fungi, plants [except blue-green algae] and animals). *Cf.* prokaryote.

eusocial Complex social organization of termites, ants and many bees and wasps, where non-reproductive individuals tend or assist reproductive individuals to whom they are closely related. See workers.

evolution Descent with modification; cumulative change in the characteristics of a population over a number of generations.

exotic species A species found outside its original geographic range (i.e., in an area where it is not native) to which it has been introduced by man either accidentally or deliberately; usually refers to animals or plants which have established self-sustaining breeding populations and hence become naturalized in the new environment; also termed **alien** species. *Cf.* native species.

fauna The animal life in a particular region, habitat or area. *Cf.* flora.

feral Referring to domestic animals which have reverted to the wild state.

filter-feeder See collector-filterer.

filtering-collector See collector-filterer.

fitness The ability of an individual to survive *and* produce viable offspring; the relative reproductive (genetic) contribution of an individual to future generations. See inclusive fitness.

flora The plant life in a particular region, habitat or area. *Cf.* fauna.

food chain An abstract model of nature which embodies the idea that organisms are dependent upon others for food and are linked together in a chain.

foraging The behaviour of an animal when it is searching for and consuming food; foraging strategy constitutes the methods employed by an animal to obtain food. See optimal foraging theory.

freshwater Water with a salinity (salt content) of less than 0.5 parts per thousand.

frugivore   An animal which eats fruit.

functional classification   Assignment of organisms into groups according to their role in a community or ecosystem, such that members of the same functional group have similar ecological roles but are not necessarily closely-related to each other in a taxonomic sense.

functional feeding groups   Classification of animals into groups according to similarity in feeding mode.

generalist   An organism occupying a broad range of habitats or eating a variety of foods. *Cf.* opportunist and specialist.

genotype   The genetic constitution of an individual. *Cf.* phenotype.

genus (pl. genera)   A category in biological classification comprising one or more morphologically-similar species; the principal category between family and species.

glacial maxima   The periods of greatest cold and farthest advance of ice sheets during the Ice Ages.

grassland   An area of vegetation dominated by grasses.

grazer-scraper   An animal which grazes the organic layer of algae, microorganisms and associated organic matter on stones and other substrates in streams; includes species which pierce plant cells and suck out the fluids.

grazer   An animals which eats plants, typically eating only part of each plant during a meal so that the effect is harmful to the plant but rarely fatal.

gross primary production   Total fixation of energy by plants in a given area per unit time. *Cf.* net primary production.

habitat   The place where an organism lives, often characterized by its physical features (e.g., the rocky hillstream habitat) or its vegetation.

herbivore   An animal which consumes living plants or their parts.

herbivory   Consumption of living plant material.

heterotrophic   Applied to organisms (animals, fungi and most bacteria) with a requirement for organic molecules as a source of energy and nutrients; also used of communities which are net importers of organic material and depend on allochthonous inputs. *Cf.* autotrophic.

hibernate   To become inactive, with a reduced metabolic rate, during the winter.

host   An organism which, often inadvertently, provides food and/or shelter for another species.

igneous   Pertaining to rocks formed by solidification of molten magma or lava.

inclusive fitness   The total fitness of an individual plus the fitness of its relatives, the latter weighted according to the degree of relatedness; i.e., devalued in proportion to their genetic difference.

indigenous   See native.

inheritance   Genetic transmission of traits (characteristics) from parent to offspring.

insectivore   An animal which eats insects.

intraspecific interactions   Interactions (such as predation or competition) occurring among members of the same species. *Cf.* interspecific interactions.

interglacial   A warm period between two glaciations or Ice Ages.

interspecific interactions   Interactions, such as predation or competition, occurring among members of the different species. *Cf.* intraspecific interactions.

Jurassic   The geological period from 210 to 140 million years ago.

keystone species   A species having a major positive or negative influence on the presence or abundance of other species; a species which has a strong influence on community structure.

labium The extensible hooked second maxillae (lower lip) of Odonata (dragonflies and damselflies) used for capturing prey; also known as a 'mask' because, when retracted, it covers the mouthparts and some of the front of the head.

Lamarckian theory The mistaken belief that characters which are acquired during the life of an organism are passed to its offspring; named after J.B.P.A. de M. Lamarck (1744–1829) who advanced an erroneous theory that evolutionary change occurred by transmission of modifications acquired through use and disuse of various organs.

larva The immature or juvenile stage of an animal where the appearance is generally unlike that of the adult; usually applied to invertebrates and lower vertebrates (amphibians, certain fishes, etc.).

larviparous Reproduction involving production of eggs which are hatched in the body of the female with release of a free-living larva; used especially of insects.

leaf area index Measure of the area of leaf surface expanded over a unit area of ground.

lignin A phenolic polymer of varied chemical composition which is a major component of the cell walls of wood and woody tissues in plants.

life cycle The sequence of stages through which an organism passes during its development from an egg to a sexually-mature adult.

litter Fallen plant parts which have not yet been decomposed fully, including leaves, wood, bark, fruit, flowers and other organic debris.

litter breakdown Loss of mass of decomposing litter, often used as a measure of litter decomposition.

litter processing Breakdown of litter brought about by animal (mostly invertebrate) feeding and microbial metabolism.

local extinction The disappearance of a species from a defined geographic area (such as Hong Kong), with the understanding that the species persists elsewhere.

longitudinal zonation   The sequential replacement of organisms, from headwaters to mouth, along a river. See zonation.

lotic   Referring to running-water habitats (rivers and streams) and their inhabitants.

macrophyte   A large (non-microscopic) plant, used especially of aquatic forms.

macroinvertebrate   An animal without a backbone that is visible to the naked eye (generally exceeding 0.5 mm body length).

microbe   See microorganism.

microhabitat   The small subset of environments within a habitat that provide suitable conditions of life for a species or a particular stage in an organism's life cycle; e.g., the stone surfaces in stream torrents where the dragonfly larva *Zygonyx iris* is found.

microorganism   Collective term for bacteria, yeasts, protozoa, blue-green algae and some fungi.

migration   Long-distance seasonal movement from one area or habitat to another. See passage migrant.

mimicry   The evolved resemblance of an organism (the mimic) to another species of organism or a non-living object. See Batesian mimicry and Müllerian mimicry.

monospecific   Made up of a single species.

Müllerian mimicry   Resemblance between two or more species of aposemete with the result that they have the same warning colouration and reinforce predator avoidance. *Cf.* Batesian mimicry.

mutualistic   Pertaining to interactions between two or more species where all parties benefit; sometimes the relationship involves complete interdependence where one species cannot survive without the other(s).

native   Occurring naturally in a particular area. *Cf.* exotic

natural selection   The process causing change in the frequency of genetically-determined characteristics of a population through differential survival and reproduction of individuals exhibiting those characteristics; the process by which organisms that are well-adapted to their environments survive, and those less well-adapted are eliminated.

naturalized   Referring to an exotic species which has established a self-sustaining population.

necrophagous   Feeding on carrion.

nectarivorous   Feeding on nectar.

nekton   Organisms which live in the water column in streams; typically they are active swimmers that can maintain their position despite the force of the current.

net primary production   Total fixation of energy by plants in a given area per unit time minus respiration over that period; gross primary production minus respiration. *Cf.* gross primary production.

neuston   Animals living on the water surface.

niche   The ecological role of a species in a community.

nuptial flight   Mating flight of some insects.

nutrient   Any substance required by an organism for growth and maintenance.

opportunist   Species which can utilize newly-available or short-lived foods or habitats, and which may switch between different types. *Cf.* generalist and specialist.

optimality theory   A theory used to predict the morphology or behaviour of an organism, based on the assumption that natural selection will favour organisms that evolve morphologies or undertake behaviour which maximize their benefit to cost ratios (or maximize net benefits) under prevailing ecological circumstances; predictions based on optimality theory are often used as a yardstick against which actual performance is assessed.

optimal foraging theory   A theory used to predict foraging behaviour on the assumption that decisions made during searching for, capturing and eating food are directed towards maximization of the net rate of food or energy intake.

organelle   Organized, discrete part inside a cell.

organic   Of biological origin (also, containing carbon).

Oriental region   A zoogeographical region, recognized by the distinctive composition of the fauna, made up of China south of the Yangtze River plus Asia south of the Himalayan-Tibetan mountain barrier (including Southeast Asia and the Indian subcontinent).

omnivore   An animal which feeds on both plants and animals.

orogeny   The process of mountain formation.

oxisol   One of the ten soil orders in a widely-used classification of soils developed by the United States Department of Agriculture. Oxisols consist largely of quartz, iron and aluminium oxides, and kaolinite clay.

oviposition   Laying eggs.

Palaearctic region   A zoogeographical region, recognized by the distinctive composition of the fauna, which includes Europe, North Africa, Siberia and those parts of Asia — including Japan — which lie north of the Yangtze River and the Himalayan-Tibetan mountain barrier.

passage migrant   A bird (or other animal) which passes through a particular area while migrating between two other areas.

patch   An area of habitat that differs with respect to its characteristics (e.g., in the density or quantity of food present) from other such areas; any habitat consists of a mosaic of patches of differing quality, value or suitability for organisms.

periphyton   The layer of algae, microorganisms and organic material coating the surface of stones, plants and hard objects on the bed of streams, rivers and reservoirs.

phenology   The study of periodic (seasonal) phenomena in plants and animals in relation to the climate.

phenotype   Physical expression of the characteristics of an organism as determined by the interaction of its genetic constitution (genotype) and the environment. *Cf.* genotype.

phenotypic plasticity   Changeability of the phenotype in response to environmental factors.

photoperiod   Duration of the period of daylight each day.

photosynthesis   Synthesis of carbohydrates from carbon dioxide and water by chlorophyll in green plants; oxygen is produced as a by-product of photosynthesis.

phylogeny   Evolutionary history of an organism.

phylum (pl. phyla)   One of the primary divisions of the animal and plant kingdoms; a group of closely-related classes of plants or animals.

plankton   Organisms which live suspended in the water.

polyvoltine   Referring to insects which can complete three or more generations per year.

population   A group of individuals of the same species in the same place at the same time.

population dynamics   Changes in the size or density of a population over time and space.

predator   An animal which kills and eats other animals (resulting in the process of predation).

primary production   Production by green plants. *Cf.* secondary production.

primary succession   A succession that takes place on a newly-exposed surface which has not borne vegetation previously.

production The quantity of organic material synthesized by organisms over a stated time period.

productivity The rate of production of organic material per unit area per unit time.

prokaryote A cell lacking a membrane-bound nucleus, as in the bacteria and blue-green algae (Cyanobacteria). *Cf.* eukaryote

prosobranch Freshwater snails which utilize dissolved oxygen with gills enclosed in a body cavity (the mantle cavity). *Cf.* pulmonate.

pulmonate Freshwater snails which utilize atmospheric oxygen with the aid of a 'lung' derived from modification of the mantle cavity. *Cf.* prosobranch.

pupa The transitional stage in the life cycle of insects that undergo complete metamorphosis during which the larval body form is reorganized to yield the final adult form.

Quaternary The geological period extending from 1.6 million years ago to the present, during which there were major climatic fluctuations (the Ice Ages).

recruitment Addition of new individuals to a population by reproduction.

resource Anything that is consumed or used up by organisms (e.g., food, nutrients, nesting sites), and hence is unavailable for other organisms and therefore potentially limiting.

rheophilic Referring to an organisms found in, and generally confined to, running waters with a swift current.

riffle Fast-flowing section of a stream where shallow water races over the stony bed.

riparian Growing or occurring along the banks of a stream or river.

River Continuum Concept The idea that running waters can be represented as a resource gradient from headwaters to mouth, and that the aquatic community is predictably structured along this resource gradient.

river regulation   General term describing modification of the natural pattern of flow of rivers and streams by extraction of water and damming, diverting or canalizing them. See channelization.

secondary production   Production by animals and microorganisms. *Cf.* primary production.

selective pressure   Any force acting on a population that results in some individuals leaving more offspring than others and so making a greater contribution to the next generation.

senescence   The process of aging.

shredder   A stream animal which feeds on litter thereby reducing it to small fragments.

shrubland   Vegetation dominated by a more-or-less complete cover of woody vegetation (especially bushes) less than 3 m in height; shrubs typically have multiple stems or trunks, whereas trees tend to have a single main trunk.

siblings   Organisms with the same parents.

Site of Special Scientific Interest (SSSI)   Areas (usually small) identified by the Hong Kong government as having special scientific or conservation value.

spate   A condition of sudden flooding.

specialist   An organism occupying one or a narrow range of habitats, or eating one or a limited range of foods. *Cf.* generalist and opportunist.

species   Can be defined in many ways, none of them entirely satisfactory. A 'biological species' is a group of organisms that can potentially interbreed to produce fertile offspring. In practice, species are recognised by their similar morphology. Such 'morphological species' usually coincide with 'biological species' in animals but often do not in plants.

spiralling   The process by which the nutrients (such as nitrogen or phosphorus) in detritus are repeatedly taken up, incorporated and then released from the bodies of stream organisms while being transported downstream by the current; measured in terms of the average distance required for one complete cycle of a nutrient, shorter distances representing 'tighter' spirals and more efficient use of nutrients contained in detrital inputs.

SSSI   See Site of Special Scientific Interest.

substratum   Any more-or-less solid surface in an aquatic environment; usually refers to the bed or bottom sediments; also termed substrate.

succession   Non-seasonal, directional change in the species composition of a community.

taxon (pl. taxa)   Any group of organisms considered to be sufficiently distinct from other such groups to be treated as a separate unit; used to generally refer to any taxonomic category of organisms, for example, nine species or nine genera could be referred to as nine taxa.

terrestrial   Living on land, as opposed to aquatic.

torpid   Dormant or inactive; in a state of torpor.

trophic   Pertaining to feeding.

trophic levels   Artificial classification of organisms according to feeding relationships and the transfer of food-energy; the first trophic level consists of primary producers (green plants); the second trophic level consists of herbivores; the third consists of animals which feed on herbivores; and so on.

trophic structure   The organization of a community in terms of energy flow through trophic levels.

turbulence   Fluid flow in which the direction and speed of movement at any point can vary rapidly.

ultisol   One of the ten soil orders in a widely-used classification of soils developed by the United States Department of Agriculture; ultisols are characterised by the accumulation of clay in the subsoil.

understorey   Forest plants which are shaded by the taller vegetation or canopy layer; the lower stratum or layer of vegetation in a forest. *Cf.* canopy.

vascular plants   Plants which possess special conducting tissues (xylem and phloem) for the transport of water and materials around the plant body; includes flowering plants, conifers, *Gnetum* and ferns.

viviparous   Reproduction involving production of active, free-living young (not eggs) from the body of the female; sometimes termed 'live-bearing'.

worker   A non-reproductive member of a eusocial insect colony which tends eggs, larvae and other individuals as well as undertaking general 'maintenance duties'; in many species, the workers are all female thus certain female wasps may forego reproduction to help their mothers produce additional, reproductive, sisters.

zonation   The spatial distribution of species along environmental gradients.

zoogeography   The study of the geographical distribution of animals and faunal groups.

Further Reading

Most of the material making up this book represents the unpublished observations and opinions of the authors. However, the manuscript was not written in an intellectual vacuum. We have benefited from the comments of colleagues and graduate students, and we have drawn freely upon published literature. In recognition of these sources, and in order to aid those who wish to find out more about the ecology of Hong Kong, we have prepared a bibliography of research papers, books and monographs that were consulted during the preparation of this book. To facilitate its use, we have given an indication of the most relevant sources for each chapter. In addition to the publications that we have drawn on, readers may find information from a variety of other sources of interest. Among these are a range of Hong Kong government publications which can be used as an aid to the identification of local flora and fauna: Johnston and Johnston (1980), Karsen *et al.* (1986), Thrower (1988) and Viney and Phillipps (1988) are particulary useful. Other publications on local natural history include the rather irregular *Memoirs of the Hong Kong Natural History Society* (starting from 1953, but originally published as *The Hong Kong Naturalist* between 1930 and 1941), the annual *Hong Kong Bird Report*, produced by the Hong Kong Bird Watching Society, and *Porcupine!*, the newsletter of The University of Hong Kong Ecology Research Group established in 1992. If you are fortunate enough to be able to locate a copy, Herklots (1951) gives an insight into natural history in Hong Kong during the first half of this century. For those seeking textbooks that explore general ecological concepts, Begon *et al.* (1990) and Stiling (1992) provide good up-to-date introductions (although they do not deal with Hong Kong) while Lincoln *et al.* (1982) serves as useful guide to the terminology of ecology and evolutionary biology.

Evolution and Adaptation

Many books have been written about evolution by natural selection and adaptation. Among the most accessible and stimulating are those authored by Richard Dawkins. In particular, Dawkins (1988) gives a clear account of the evidence for evolution, while Dawkins (1989) discusses the 'selfish gene', and its implications for behaviour and ecology, in terms of the theory of natural selection.

Environment and History

Data on the climate and seasonality of Hong Kong are available in the monthly weather summary published by the Royal Observatory (Hong Kong government). In addition to a summary of the month's weather, these reports include tables of monthly means of various climatic parameters for the previous 30 years and the extreme values for the same parameters since records began in 1884. The December issue for each year also includes a summary of the year's weather. Chin (1986) gives a more general account of seasonal variations in Hong Kong's weather. Jayawardena and Peart (1989) have described the effect that seasonal rainfall has on water discharge volumes in Hong Kong streams.

Our description of the geology of Hong Kong is based largely on information given by McFeat-Smith *et al.* (1989), while our account of Hong Kong soils is derived from that of Grant (1986).

The complete history of human impacts upon the Hong Kong environment will never be known. Most historians have devoted their energies to documenting events since 1842 when Hong Kong became a British possession (e.g., Endacott 1973; Cameron 1991) and, indeed, Endacott (1973) stated that '. . . the history of Hong Kong really begins with the coming of the British in 1841'. Human impacts upon the environment, however, started long before the arrival of the British, although only fragmentary information on their nature and extent is available. Ng (1983) provides a summary and extracts of the Chinese gazetteers for the Hong Kong region. Accounts of clan lineages and past agricultural practices in the New Territories give some indication of likely human impacts upon natural habitats, and in this regard we have found articles or books by Gibbs (1931a), Sung (1935a, 1935b), Hayes (1977, 1983, 1986), Faure (1984, 1986), Faure *et al.* (1984) and Siu (1984) particularly helpful. A lot of useful information on post-1841 impacts is buried in the annual reports of the various government departments which have been responsible for forestry and agriculture, from the Botanical and Afforestation Department of the 1880s to the Agriculture and Fisheries Department of today. Records of visits by European naturalists in the nineteenth and early twentieth centuries (Hinds and Bentham 1842; Seeman 1852; Swinhoe 1861; Lockhart 1898; Kershaw 1904) also contain insights into past human impacts. Archaeological investigations have provided evidence for Stone Age settlements dating back at least 6000 years and, as new sites are investigated, we will undoubtedly learn more about human impacts during this period. Accounts of these investigations appear in the local newspapers and the *Journal of the Hong Kong Archaeological Society*.

Climate and the Hong Kong Biota

As mentioned above, information on the weather in any particular year, and its relationship to long-term averages of various meteorological parameters, is available from reports compiled by the Royal Observatory. In addition, there are reports dealing with the effects of extreme weather conditions in Hong Kong, in particular, the great frost of 1893 (Ford *et al.* 1893; Skertchly 1893; Gibbs 1931b). Some data relating to more recent cold snaps have been published also (Bannister 1948; Peart and Guan 1992), but only Corlett (1992a) has directly addressed the effects of extreme weather conditions on the Hong Kong biota. We have used these data, combined with our knowledge of the local flora and fauna and the distribution of these species beyond the Territory (e.g., Allen 1938; Corbet and Hill 1992; Zhao and Adler 1993), to infer the influence of climate and geography on the composition of the Hong Kong biota. Our general conclusions agree with those arising from a quantitative biogeographic analysis of Hong Kong caddisflies (Dudgeon 1987a).

Seasonality

A good deal is known about the seasonality of the Hong Kong biota and much useful data on the leafing, fruiting and flowering periods of local plants, the breeding periods of animals, and the timing of bird migration is included in guides to the local flora and fauna, in particular, Marshall (1967), Karsen *et al.* (1986), Thrower (1988), Viney and Phillipps (1988) and Ades (1990). A detailed study of the seasonality of the local shrubland flora has been undertaken by Corlett (1993), while the pattern of leaf litter fall in secondary forest has been documented by Dudgeon (1982a) and Lam and Dudgeon (1985a). Changes in the abundance and species diversity of carrion flies and woodlice (isopods) in Hong Kong forests are described by So and Dudgeon (1990a) and Ma *et al.* (1991a, 1991b). Lam and Dudgeon (1984) have made a preliminary comparison of the pattern of seasonal changes in the abundance of ground-dwelling insects in forest, shrubland and grassland.

There is a relatively large amount of information on the seasonality of aquatic animals and freshwater habitats. Dudgeon (1988a, 1992a) has summarized the seasonal changes which occur in local streams and reservoirs, while a number of investigations of the population dynamics, reproduction and other aspects of the seasonality of aquatic molluscs (Dudgeon 1982b, 1983, 1986; Dudgeon and Morton 1983, 1984; Morton 1977a, 1977b, 1983a, 1985, 1986, 1989a), crustaceans (Dudgeon 1985, 1987b; Ng and Dudgeon 1992) and insects (Dudgeon 1984a, 1988b, 1989b, 1989c, 1990a; Dudgeon and Wat 1986; Wells and Dudgeon 1990) have been undertaken. Karsen *et al.* (1986) have documented the timing of breeding by freshwater reptiles and amphibians, and Dudgeon (1992a) gives some preliminary data on fishes.

Succession and Climax

Much of our knowledge of the present-day vegetation of Hong Kong is the result of field surveys carried out by one of us (Corlett, unpublished) and by our graduate students. World Wide Fund for Nature (Hong Kong), in conjunction with the Survey and Mapping Office of the Hong Kong Government, have recently produced a

1:50 000 scale vegetation map of Hong Kong, together with an explanatory booklet (Ashworth *et al.* 1993). References to historical impacts on the vegetation have been mentioned above (see Environment and History). Research on frugivory and seed dispersal by animals is continuing and preliminary results have been published (Corlett 1992b, 1992c). While more data are needed, it appears that many aspects of plant-frugivore interactions in Hong Kong are similar to those elsewhere in tropical Asia (Corlett and Lucas 1990; Phua and Corlett 1989).

Land and Water

The abiotic characteristics of Hong Kong streams and the influence that they have on distribution and the zonation of the fauna, have been described by Dudgeon (1982c, 1982d, 1982e, 1983a, 1984a, 1987c, 1989b, 1989d, 1990b). The importance of leaf litter and terrestrial influences on stream ecology in Hong Kong is discussed by Dudgeon (1982a, 1982f, 1983b, 1987d, 1988c, 1992a). While Hynes (1975) was perhaps the first to draw attention to the close links between streams and their valleys, this idea has now been embodied in the River Continuum Concept (Vannote *et al.* 1980) which has been applied to Hong Kong streams by Dudgeon (1984b, 1989e). It is now apparent that land-water interactions not only exert a dominating influence on the ecology of streams in Hong Kong and elsewhere (Dudgeon 1984c, 1991a), but also have important effects upon reservoirs such as Plover Cove in Hong Kong (Dudgeon 1983c, 1983d, 1987e). A knowledge of the magnitude and extent of land-water interactions may provide a basis for conservation and management strategies for of freshwater habitats (Dudgeon 1987f, 1991a, 1992b).

Foods and Feeding

Information on primary production and herbivory in terrestrial habitats in Hong Kong is scarce, although Lam and Dudgeon (1985b) have discussed the implications of herbivory for one local fig species. There are more data on primary production and herbivory in local streams. Dudgeon (1983b) has estimated primary production by algae in Tai Po Kau Forest Stream, and Dudgeon and Chan (1992) have documented the effects of variations in algal standing stocks on the distribution and abundance of herbivores in streams. The diets of some freshwater herbivores are described by Dudgeon (1983d, 1987c) and Dudgeon and Yipp (1985; see also Dudgeon [1992a]). Detritivory and the importance of animals in leaf-litter breakdown in Hong Kong forests has been addressed by Lam and Dudgeon (1985c), Dudgeon *et al.* (1990) and Lam *et al.* (1981). Dudgeon (1982f) has investigated the litter breakdown in Tai Po Kau Forest Stream and upper Lam Tsuen River with particular emphasis on the role played by detritivorous invertebrates. Preliminary data of the effects of detritus on the distribution and abundance of stream insects is given by Dudgeon (1990b, 1993), while the importance of algae versus detritus in structuring stream communities is discussed by Dudgeon (1988c, 1989e).

Data on frugivory and seed predation by terrestrial vertebrates is given by Corlett (1992b, 1992c) and Fellowes (1992). A great deal more is known but still unpublished. The terrestrial invertebrates associated with rotting fruit have been studied by Lam and Dudgeon (1983, 1984) who also discuss the habits of some of insects attracted to carrion. This point has been elaborated in more detail by So and Dudgeon (1989a, 1989b, 1990a, 1990b), with particular reference to outcome of competition for food among the larvae of different types of carrion flies.

Information on the diet of mammalian carnivores has been collated from examination of scats and from unpublished field observations made by ourselves, our graduate students and by local naturalists. Little has been published on prey defense mechanisms in Hong Kong, although see Edmunds (1974) for a fascinating general survey of anti-predator devices. However, Johnston and Johnston (1980) have illustrated some of the mimetic butterflies found locally, and Edmunds and Dudgeon (1981) give a brief account of crypsis in a Hong Kong praying mantis. Carnivory in streams has been investigated largely by means of gut-content analysis of predators (Dudgeon 1987c, 1989b, 1989d), although field experiments have been employed to estimate the impact of predatory fish on stream invertebrate communities (Dudgeon 1991c, 1993).

The basic tenets and predictions of optimal foraging theory have been reviewed by Pyke *et al.* (1977), Pyke (1984) and Krebs and McCleery (1984), while Krebs and Davies (1993, especially Chapter 3) give a readable and up-to-date account of the main issues. The foraging behaviour of Hong Kong freshwater snails has been described by Dudgeon and Lam (1985a, 1985b), while Dudgeon and Cheung (1990) and Dudgeon (1990c) have used predictions derived from optimal foraging theory to investigate prey selection by freshwater predators.

Aliens

Many species of plants and animals have been introduced to Hong Kong and have managed to establish themselves. Early observations on introduced plants were made by Crook (1931, 1932), while Corlett (1992d) gives a recent account of the composition of the naturalized exotic flora of Hong Kong and notes that new introductions are still taking place (Corlett 1992e). Fellowes (1992) has undertaken a detailed study of the introduced Hong Kong macaques, and documents their population growth and interactions with man. Jarrett (1931) has described the early success of introduced African land snails, while Winney (1983) summarized information on the pinewood nematode in Hong Kong and its devastating effects on the on Chinese red pine. Data on aliens in freshwaters — in particular, those habitats which have been disturbed by human activities — is given by Morton (1975), Dudgeon and Yipp (1983), Yipp (1990, 1991), Yipp *et al.* (1992), Chong and Dudgeon (1992) and Dudgeon (1992a, 1992c).

Conservation

Thrower (1984) provides a useful guide to the Hong Kong Country Parks system but, unfortunately, this publication is out of print. At the time of writing there has been no overall conservation evaluation of Hong Kong, nor has there been any systematic attempt to identify priority habitats with respect to their conservation value. Work is now under way to rectify this situation, and the World Wide Fund for Nature (Hong Kong), in collaboration with local naturalists and academics, is progressing towards production of a territory-wide ecological map including vegetation types, priority habitats, SSSIs, etc. (Ashworth *et al.* 1993). The laws relating to nature conservation are included under a number of bills and ordinances, and no review is available at present. The Environmental Protection Department (Hong

Kong Government) produces annual reports on the state of Hong Kong's environment (e.g., EPD 1993), with particular emphasis on control measures, but they do not deal with nature conservation issues. Of particular concern is the fact that there is little information on the conservation status of particular groups or species of plants and animals (see Chong and Dudgeon 1992), although some publications (e.g., Karsen *et al*. 1986; Viney and Phillipps 1988; Ades 1990) do give an indication of the commonness or rarity of particular species.

Bibliography

Ades, G. 1990. *Bats of Hong Kong*. Hong Kong: World Wide Fund for Nature.

Allen, G.M. 1938. *The Mammals of China and Mongolia*. Natural History of Central Asia Series, Vol. XI (parts 1 and 2). New York: American Museum of Natural History.

Ashworth, J.M., Corlett, R.T., Dudgeon, D., Melville, D.S. and Tang, W.S.M. 1993. *Hong Kong Flora and Fauna: Computing Conservation. Hong Kong Ecological Database*. Hong Kong: World Wide Fund for Nature.

Bannister, R.C. 1948. A remarkable cold spell at Hong Kong. *Weather* 3: 344.

Begon, M., Harper, J.L. and Townsend, C.R. 1990. *Ecology: Individuals, Populations and Communities*. 2nd ed. Oxford: Blackwell Scientific Publishers.

Cameron, N. 1991. *An Illustrated History of Hong Kong*. Hong Kong: Oxford University Press.

Chin, P.C. 1986. Climate and weather. In *A Geography of Hong Kong*. 2nd ed. (T.N. Chiu and C.L. So, eds.), 69–85. Hong Kong: Oxford University Press.

Chong, D.H. and Dudgeon, D. 1992. Hong Kong stream fishes: an annotated checklist with remarks on conservation status. *Memoirs of the Hong Kong Natural History Society* 19: 79–112.

Corbet, G.B. and Hill, J.E. 1992. *The Mammals of the Indomalayan Region*. Oxford: Oxford University Press.

Corlett, R.T. 1992a. The impact of cold and frost on terrestrial vegetation in Hong Kong. *Memoirs of the Hong Kong Natural History Society* 19: 133–5.

Corlett, R.T. 1992b. Plants attractive to frugivorous birds in Hong Kong. *Memoirs of the Hong Kong Natural History Society* 19: 115–6.

Corlett, R.T. 1992c. Seed dispersal by birds in Hong Kong shrubland. *Memoirs of the Hong Kong Natural History Society* 19: 129–30.

Corlett, R.T. 1992d. The naturalized flora of Hong Kong. *Journal of Biogeography* 19: 421–30.

Corlett, R.T. 1992e. Two additions to the naturalized flora of Hong Kong. *Memoirs of the Hong Kong Natural History Society* 19: 119.

Corlett, R.T. 1993. The reproductive phenology of Hong Kong shrubland. *Journal of Tropical Ecology* 9: 501–10.

Corlett, R.T. and Lucas, P.W. 1990. Alternative seed-handling strategies in primates: seed-spitting by Long-tailed Macaques (*Macaca fascicularis*). *Oecologia* 82: 166–71.

Crook, A.H. 1931. Some visitors which came to stay. *The Hong Kong Naturalist* 2: 178–84.

Crook, A.H. 1932. Some visitors which came to stay, Part II. *The Hong Kong Naturalist* 3: 11–5.

Dawkins, R. 1988. *The Blind Watchmaker*. London: Penguin Books.

Dawkins, R. 1989. *The Selfish Gene*. 2nd ed. Oxford: Oxford University Press.

Dudgeon, D. 1982a. Spatial and seasonal variations in the standing crop of periphyton and allochthonous detritus in a forest stream in Hong Kong, with notes on the magnitude and fate of riparian leaf fall. *Archiv für Hydrobiologie-Supplement* 64: 189–220.

Dudgeon, D. 1982b. The life history of *Brotia hainanensis* (Brot, 1872) (Gastropoda: Prosobranchia: Thiaridae) in a tropical forest stream. *Zoological Journal of the Linnean Society* 76: 141–54.

Dudgeon, D. 1982c. Aspects of the hydrology of Tai Po Kau Forest Stream, New Territories, Hong Kong. *Archiv für Hydrobiologie-Supplement* 64: 1–35.

Dudgeon, D. 1982d. Spatial and temporal changes in the sediment characteristics of Tai Po Kau Forest Stream, New Territories, Hong Kong, with some preliminary observations upon within-reach variations in current velocity. *Archiv für Hydrobiologie-Supplement* 64: 36–63.

Dudgeon, D. 1982e. Aspects of the microdistribution of insect macrobenthos in a forest stream in Hong Kong. *Archiv für Hydrobiologie-Supplement* 64: 267–71, 1982.

Dudgeon, D. 1982f. An investigation of physical and biological processing of two species of leaf litter in Tai Po Kau Forest Stream, New Territories, Hong Kong. *Archiv für Hydrobiologie* 96: 1–32.

Dudgeon, D. 1983a. Spatial and temporal changes in the distribution of gastropods in the Lam Tsuen River, New Territories, Hong Kong, with notes on the occurrence of the exotic snail, *Biomphalaria straminea*. *Malacological Review* 16: 91–2.

Dudgeon, D. 1983b. Preliminary measurements of primary production and community respiration in a forest stream in Hong Kong. *Archiv für Hydrobiologie* 98: 287–98.

Dudgeon, D. 1983c. The effects of water level fluctuations on a gently shelving marginal zone of Plover Cove Reservoir, Hong Kong. *Archiv für Hydrobiologie-Supplement* 65: 163–96.

Dudgeon, D. 1983d. The utilization of terrestrial plants as a food source by the fish stock of a gently sloping marginal zone in Plover Cove Reservoir, Hong Kong. *Environmental Biology of Fishes* 8: 73–7.

Dudgeon, D. 1984a. Seasonal and long-term changes in the hydrobiology of the Lam Tsuen River, New Territories, Hong Kong, with special reference to benthic macroinvertebrate distribution and abundance. *Archiv für Hydrobiologie-Supplement* 69: 55–129.

Dudgeon, D. 1984b. Longitudinal and temporal changes in macroinvertebrate community organization in the Lam Tsuen River, Hong Kong. *Hydrobiologia* 111: 207–17.

Dudgeon, D. 1984c. The importance of streams in tropical rain-forest systems. In *Tropical Rain-forest: The Leeds Symposium* (A.C. Chadwick and S.L. Sutton, eds.), 71–82. Leeds: Special Publication of the Leeds Literary and Philosophical Society (Scientific Section).

Dudgeon, D. 1985. The population dynamics of some freshwater carideans (Crustacea: Decapoda) in Hong Kong, with special reference of *Neocaridina serrata* (Atyidae). *Hydrobiologia* 120: 141–9.

Dudgeon, D. 1986. The life cycle, population dynamics and productivity of *Melanoides tuberculata* (Müller, 1774) (Gastropoda: Prosobranchia: Thiaridae) in Hong Kong. *Journal of Zoology, London* 208: 37–53.

Dudgeon, D. 1987a. Preliminary investigations on the faunistics and ecology of Hong Kong Trichoptera. In *Proceedings of the 5th International Symposium on Trichoptera* (M. Bournaud and H.

Tachet, eds.), 111–7. Series Entomologica. The Hague: Dr W. Junk Publishers.

Dudgeon, D. 1987b. The larval development of *Neocaridina serrata* (Stimpson) (Crustacea: Caridea: Atyidae) from Hong Kong. *Archiv für Hydrobiologie* 110: 339–56.

Dudgeon, D. 1987c. Niche specificities of four fish species (Homalopteridae, Cobitidae and Gobiidae) from a Hong Kong forest stream. *Archiv für Hydrobiologie* 108: 349–64.

Dudgeon, D. 1987d. The ecology of a forest stream in Hong Kong. *Archiv für Hydrobiologie Beiheft — Ergebnisse der Limnologie* 28: 449–54.

Dudgeon, D. 1987e. The development of benthic macroinvertebrate communities in Plover Cove Reservoir, Hong Kong, with particular reference to the significance of the marginal zone. *Archiv für Hydrobiologie Beiheft — Ergebnisse der Limnologie* 28: 497–502.

Dudgeon, D. 1987f. Three contrasting land-water interactive systems in Hong Kong. *Archiv für Hydrobiologie Beiheft — Ergebnisse der Limnologie* 28: 417–20.

Dudgeon, D. 1988a. Hong Kong freshwaters: seasonal influences on benthic communities. *Verhandlungen Internationale Vereinigung für theoretische und angewandte Limnologie* 23: 1362–6.

Dudgeon, D. 1988b. Flight periods of aquatic insects from a Hong Kong forest stream I. Macronematinae (Hydropsychidae) and Stenopsychidae (Trichoptera). *Aquatic Insects* 10: 61–8.

Dudgeon, D. 1988c. The influence of riparian vegetation on macroinvertebrate community structure in four Hong Kong streams. *Journal of Zoology, London* 216: 609–27.

Dudgeon, D. 1989a. Ecological strategies of Hong Kong Thiaridae (Gastropoda: Prosobranchia). *Malacological Review* 22: 39–53.

Dudgeon, D. 1989b. Life cycle, production, microdistribution and diet of the damselfly *Euphaea decorata* (Odonata: Euphaeidae) in a Hong Kong forest stream. *Journal of Zoology, London* 217: 57–72.

Dudgeon, D. 1989c. Gomphid (Odonata: Anisoptera) life cycles and production in a Hong Kong forest stream. *Archiv für Hydrobiologie* 114: 531–6.

Dudgeon, D. 1989d. Resource partitioning among Odonata (Insecta: Anisoptera and Zygoptera) larvae in a Hong Kong forest stream. *Journal of Zoology, London* 217: 381–402.

Dudgeon, D. 1989e. The influence of riparian vegetation on the functional organization of four Hong Kong stream communities. *Hydrobiologia* 179: 183–94.

Dudgeon, D. 1990a. Seasonal dynamics of drift in a Hong Kong stream. *Journal of Zoology, London* 222: 187–96.

Dudgeon, D. 1990b. Determinants of the distribution and abundance of larval Ephemeroptera (Insecta) in Hong Kong running waters. In *Mayflies and Stoneflies: Biology and Life Histories* (I.C. Campbell, ed.), 221–32. Dordrecht: Kluwer Academic Publishers.

Dudgeon, D. 1990c. Feeding by the aquatic heteropteran *Diplonychus rusticum*: an effect of prey density on meal size. *Hydrobiologia* 190: 93–6.

Dudgeon, D. 1991a. The stream and its valley: human interference with fluvial ecosystems. In: *Polmet '91: Pollution in the Metropolitan and Urban Environment* (J. Boxall, ed.), 823–38. Hong Kong: Hong Kong Institution of Engineers.

Dudgeon, D. 1991b. An experimental study of abiotic disturbance effects on community structure and function in a tropical stream. *Archiv für Hydrobiologie* 122: 403–20.

Dudgeon, D. 1991c. An experimental study of the effects of predatory fish on macroinvertebrates in a Hong Kong stream. *Freshwater Biology* 25: 321–30.

Dudgeon, D. 1992a. *Patterns and Processes in Stream Ecology.* Stuttgart: Schweizerbart'sche Verlagsbuchhandlung.

Dudgeon, D. 1992b. Endangered ecosystems: a review of the conservation status of tropical Asian rivers. *Hydrobiologia* 234: 1–25.

Dudgeon, D. 1992c. The effects of water transfer on aquatic insects in a Hong Kong stream. *Regulated Rivers: Research and Management* 7: 369–77.

Dudgeon, D. 1993. The effects of spate-induced disturbance, predation and environmental complexity on macroinvertebrates in a tropical stream. *Freshwater Biology* 30: 189–97.

Dudgeon, D. and Chan, I.K.K. 1992. An experimental study of the influence of periphytic algae on invertebrate abundance in a Hong Kong stream. *Freshwater Biology* 27: 53–63.

Dudgeon, D. and Cheung, C.P.S. 1990. Selection of gastropod prey by a tropical freshwater crab. *Journal of Zoology, London* 220: 147–55.

Dudgeon, D. and Lam, P.K.S. 1985a. Freshwater gastropod foraging strategies: interspecific comparisons. In *The Malacofauna of Hong Kong and Southern China II* (B. Morton and D. Dudgeon, eds.), 591–600. Hong Kong: Hong Kong University Press.

Dudgeon, D. and Lam, P.K.S. 1985b. The effects of feeding and starvation on the foraging strategies of freshwater pulmonates. In *The Malacofauna of Hong Kong and Southern China II* (B. Morton and D. Dudgeon, eds.), 601–12. Hong Kong: Hong Kong University Press.

Dudgeon, D. and Morton, B. 1983. The population dynamics and sexual strategy of *Anodonta woodiana* (Bivalvia: Unionacea) in Plover Cove Reservoir, Hong Kong. *Journal of Zoology, London* 201: 161–83.

Dudgeon, D. and Morton, B. 1984. Site selection and attachment duration of *Anodonta woodiana* (Bivalvia: Unionacea) glochidia on fish hosts. *Journal of Zoology, London* 204: 355–62.

Dudgeon, D. and Wat, C.Y.M. 1986. Life cycle and diet of *Zygonyx iris insignis* (Insecta: Odonata: Anisoptera) in Hong Kong running waters. *Journal of Tropical Ecology* 2: 75–87.

Dudgeon, D. and Yipp, M.W. 1983. A report on the gastropod fauna of aquarium fish farms in Hong Kong, with special reference to an introduced human schistosome host species, *Biomphalaria straminea* (Pulmonata: Planorbidae). *Malacological Review* 16: 93–4.

Dudgeon, D. and Yipp, M.W. 1985. The diets of Hong Kong freshwater gastropods. In *The Malacofauna of Hong Kong and Southern China II* (B. Morton and D. Dudgeon, eds.), 491–509. Hong Kong: Hong Kong University Press.

Dudgeon, D., Lam, P.K.S. and Ma, H.T. 1990. Differential palatability of leaf litter to four sympatric isopods in a Hong Kong forest. *Oecologia (Berlin)* 84: 398–403.

Edmunds, M. 1974. *Defense in Animals: A Survey of Anti-predator Defenses*. Harlow: Longman.

Edmunds, M. and Dudgeon, D. 1991. Cryptic behaviour in the Oriental leaf mantis *Sinomantis denticulata* Beier (Dictyoptera: Mantodea). *Entomologists' Monthly Magazine* 127: 45–8.

Endacott, G.B. 1973. *A History of Hong Kong*. Hong Kong: Oxford University Press.

EPD. 1993. *Environment Hong Kong 1993: A Review of 1992*. Hong Kong: Environmental Protection Department, Government Printer.

Faure, D. 1984. Lineage village and alliance: the territorial organization of the New Territories. *Proceedings of the Sixth International Symposium on Asian Studies, 1984*, 543–54. Hong Kong: Asian Research Services.

Faure, D. 1986. *The Structure of Chinese Rural Society: Lineage and Village in the Eastern New Territories, Hong Kong*. Hong Kong: Oxford University Press.

Faure, D., Luk, K.K.B. and Ng, N.L.A. 1984. The Hong Kong region according to historical inscriptions. In *From Village to City: Studies in the Traditional Roots of Hong Kong Society* (D. Faure, J. Hayes and A. Birch, eds.), 43–54. Hong Kong: Centre of Asian Studies, University of Hong Kong.

Fellowes, J.F. 1992. *Hong Kong Macaques*. Final Report to the WWF Hong Kong Projects Committee, World Wide Fund for Nature, Hong Kong.

Ford, C., Thiselton-Dyer, W.T. and Doberck, W. 1893. Severe frost at Hong Kong. *Nature (London)* 47 (1223): 535–6.

Gibbs, L. 1931a. Agriculture in the New Territory. *The Hong Kong Naturalist* 2: 132–4.

Gibbs, L. 1931b. The Hong Kong frost of January 1893. *The Hong Kong Naturalist* 2: 318–9.

Grant, C.J. 1986. Soil. In A *Geography of Hong Kong*. 2nd ed. (T.N. Chiu and C.L. So, eds.), 110–7. Hong Kong: Oxford University Press.

Hayes, J. 1977. *The Hong Kong Region 1850–1911. Institutions and Leadership in Town and Countryside*. Archon Books/Dawson, Hamden, U.S.A. and Folkstone, England.

Hayes, J. 1983. *The Rural Communities of Hong Kong: Studies and Themes*. Hong Kong: Oxford University Press.

Hayes, J. 1986. Stakenet and fishing canoe: Hong Kong and adjacent islands in the 19th and early 20th centuries. The sea and the shore in social, economic and political organization. *Proceedings of the Eighth International Symposium on Asian Studies, 1986*, 573–98. Hong Kong: Asian Research Service.

Herklots, G.A.C. 1951. *The Hong Kong Countryside*. Hong Kong: South China Morning Post.

Hinds, R.B. and Bentham, G. 1842. Remarks on the physical aspect, climate, and vegetation of Hong-Kong, China, with an enumeration of plants there collected. *London Journal of Botany* 1: 476–81.

Hynes, H.B.N. 1975. The stream and its valley. *Verhandlungen Internationale Vereinigung für theoretische und angewandte Limnologie* 19: 1–15.

Jarrett, V.H.C. 1931. The spread of the snail *Achatina fulica* in South China. *The Hong Kong Naturalist* 2: 262–4.

Jayawardena, A.W. and Peart, M.R. 1989. Spatial and temporal variation of rainfall and runoff in Hong Kong. In *FRIENDS in Hydrology* (L. Roald, K. Nordseth and K. Anker-Hassel, eds.), 409–18. IAHS (International Association of Hydrological Sciences) Publication No. 187. Wallingford: IAHS Press.

Johnston, G. and Johnston, B. 1980. *This is Hong Kong: Butterflies.* Hong Kong: Information Services Department.

Karsen, S.J., Lau, M.W. and Bogadek, A. 1986. *Hong Kong Amphibians and Reptiles.* Hong Kong: An Urban Council Publication.

Kershaw, J.C. 1904. List of the birds of the Quangtung coast, China. *Ibis* 4: 235–48.

Krebs, J.R. and Davies, N.B. 1993. *An Introduction to Behavioural Ecology.* 2nd ed. Oxford: Blackwell Scientific Publications.

Krebs, J.R. and McCleery, R.H. 1984. Optimization in behavioural ecology. In *Behavioural Ecology: an Evolutionary Approach.* 2nd ed. (J.R. Krebs and N.B. Davies, eds.), 91–121. Oxford: Blackwell Scientific Publications, Oxford.

Lam, P.K.S. and Dudgeon, D. 1983. Ecological studies of necrophilous and carpophilous beetles in three different habitats in Hong Kong. I. Spatial variations. *Ecologic Science* 1983 (2): 24–33.

Lam, P.K.S. and Dudgeon, D. 1984. Ecological studies of necrophilous and carpophilous beetles in three different habitats in Hong Kong. II. Temporal variations. *Ecologic Science* 1984 (1): 86–97.

Lam, P.K.S. and Dudgeon, D. 1985a. Seasonal effects on litterfall in a Hong Kong mixed forest. *Journal of Tropical Ecology* 1: 55–64.

Lam, P.K.S. and Dudgeon, D. 1985b. The effects and possible implications of artificial damage on the life-span of *Ficus fistulosa* leaves. *Journal of Tropical Ecology* 1: 187–90.

Lam, P.K.S. and Dudgeon, D. 1985c. Breakdown of *Ficus fistulosa* (Moraceae) leaves in Hong Kong, with special reference to dynamics of elements and the effects of invertebrate consumers. *Journal of Tropical Ecology* 1: 249–64.

Lam, P.K.S., Dudgeon, D. and Ma, H.T. 1991. Ecological energetics of populations of four sympatric isopods in a Hong Kong forest. *Journal of Tropical Ecology* 7: 475–90.

Lincoln, R.J., Boxhall, G.A. and Clark, P.F. 1983). *A Dictionary of Ecology, Evolution and Systematics.* Cambridge: Cambridge University Press.

Lockhart, S. 1898. *Extracts from a Report by Mr. Stewart Lockhart on the Extension of the Colony of Hong Kong.* Paper laid before the Legislative Council of Hong Kong, 1899.

Ma, H.T., Dudgeon, D. and Lam, P.K.S. 1991a. Seasonal changes in populations of three sympatric isopods in a Hong Kong forest. *Journal of Zoology, London* 224: 347–65.

Ma, H.T., Lam, P.K.S. and Dudgeon, D. 1991b. Inter- and intraspecific life-history variations of three sympatric isopods in a Hong Kong forest. *Journal of Zoology, London* 224: 667–87.

Marshall, P. 1967. *Wild Mammals of Hong Kong*. Hong Kong: Oxford University Press.

McFeat-Smith, I., Workman, D.R., Burnett, A.D. and Chau, E.P.Y. 1989. Geology of Hong Kong. *Bulletin of the Society of Engineering Geologists* 26: 23–107.

Morton, B. 1975. The colonization of Hong Kong's raw water supply system by *Limnoperna fortunei* (Dunker) (Bivalvia: Mytilaceae) from China. *Malacological Review* 8: 91–105.

Morton, B. 1977a. The population dynamics of *Limnoperna fortunei* (Dunker 1857) (Bivalvia: Mytilaceae) in Plover Cove Reservoir, Hong Kong. *Malacologia* 16: 165–82.

Morton, B. 1977b. The population dynamics of *Corbicula fluminea* (Müller 1774) (Bivalvia: Corbiculaceae) in Plover Cove Reservoir, Hong Kong. *Journal of Zoology, London* 181: 21–42.

Morton, B. 1983. The sexuality of *Corbicula fluminea* (Bivalvia: Corbiculaceae) in lentic and lotic waters in Hong Kong. *Journal of Molluscan Studies* 49: 81–3.

Morton, B. 1985. The population dynamics, reproductive strategy and life history of *Musculium lacustre* (Pisidiidae) in Hong Kong. *Journal of Zoology, London* 207: 581–603.

Morton, B. 1986. The population dynamics and life history tactics of *Pisidium clarkeanum* and *P. annandalei* (Bivalvia: Pisidiidae) sympatric in Hong Kong. *Journal of Zoology, London* 210: 427–49.

Ng, P.K.L. and Dudgeon, D. 1992. The Potamidae and Parathelphusidae (Crustacea: Decapoda: Brachyura) of Hong Kong. *Invertebrate Taxonomy* 6: 741–68.

Ng, P.Y.L. 1983. *New Peace County: A Chinese Gazetteer of the Hong Kong Region*. Hong Kong: Hong Kong University Press.

Peart, M.R. and Gaun, D. 1992. A cold spell in Hong Kong. *Hong Kong Meteorological Society Bulletin* 2: 12–7.

Phua, P.B. and Corlett, R.T. 1989. Seed dispersal by the Lesser Short-nosed Fruit Bat (*Cynopterus brachyotis*, Pteropodidae, Megachiroptera). *Malayan Nature Journal* 42: 251–6.

Pyke, G.H. 1984. Optimal foraging theory: a critical review. *Annual Review of Ecology and Systematics* 15 523–75.

Pyke, G.H., Pulliam, H.R. and Charnov, E.L. 1977. Optimal foraging: a selective review of theory and tests. *Quarterly Review of Biology* 52: 137–54.

Seeman, B. 1852. *The Botany of the Voyage of H.M.S. Herald, under the Command of Captain Henry Kellet, R.N., C.B., during the Years 1845–51.* London: Lovell Reeve.

Siu, K.A. 1984. The Hong Kong region before and after the coastal evacuation in the early Ch'ing Dynasty. In *From Village to City: Studies in the Traditional Roots of Hong Kong Society* (D. Faure, J. Hayes and A. Birch, eds.), 1–9. Hong Kong: Centre of Asian Studies, University of Hong Kong.

Skertchly, S.B.J. 1893. The cold wave at Hongkong, January 1893. — Its after effects. *Nature (London)* 48 (1227): 3–5.

So, P.M. and Dudgeon, D. 1989a. Variations in the life-history parameters of *Hemipyrellia ligurriens* (Diptera: Calliphoridae) in response to larval competition for food. *Ecological Entomology* 14: 109–16.

So, P.M. and Dudgeon, D. 1989b. Life-history responses of larviparous *Boettcherisca formosensis* (Diptera: Sarcophagidae) to larval competition for food, including a comparison with oviparous *Hemipyrellia ligurriens* (Calliphoridae). *Ecological Entomology* 14: 349–56.

So, P.M. and Dudgeon, D. 1990a. Phenology and diversity of necrophagous Diptera in a Hong Kong forest. *Journal of Tropical Ecology* 6: 91–101.

So, P.M. and Dudgeon, D. 1990b. Interspecific competition among larvae of *Hemipyrellia ligurriens* (Calliphoridae) and *Boettcherisca formosensis* (Sarcophagidae) (Diptera). *Researches on Population Ecology (Kyoto)* 32: 337–48.

Stiling, P.D. 1992. *Introductory Ecology.* New Jersey: Prentice Hall Inc.

Sung, H. 1935a. Legends and stories of the New Territories. I. Tai Po. *The Hong Kong Naturalist* 6: 36–9.

Sung, H. 1935b. Legends and stories of the New Territories. III. Kam Tin. *The Hong Kong Naturalist* 6: 213–8.

Swinhoe, R. 1861. Notes on the ornithology of Hong Kong, Macao and Canton, made during the latter end of February March, April, and the beginning of May, 1860. *Ibis* 1861: 23–57.

Thrower, S.L. 1984. *Hong Kong Country Parks.* Hong Kong: Hong Kong Government Printer.

Thrower, S.L. 1988. *Hong Kong Trees (Omnibus Volume).* Hong Kong: Urban Council Publications.

Vannote, R.L., Minshall, G.W., Cummins, K.W., Sedell, J.R. and Cushing, C.E. 1980. The River Continuum Concept. *Canadian Journal of Fisheries and Aquatic Sciences* 37: 130–7.

Viney, C. and Phillipps, K. 1988. *Birds of Hong Kong*. 4th ed. Hong Kong: Hong Kong Government Printer.

Wells, A. and Dudgeon, D. 1990. Hydroptilidae (Insecta) from Hong Kong. *Aquatic Insects* 12: 161–75.

Winney, R. 1983. Preliminary observations on the Pinewood Nematode *Busaphelenchus xylophilus* in Hong Kong. Unpublished report of the Hong Kong Agriculture and Fisheries Department, Hong Kong Government.

Yipp, M.W. 1990. Distribution of the schistosome vector snail *Biomphalaria straminea* (Pulmonata: Planorbidae) in Hong Kong. *Journal of Molluscan Studies* 56: 47–55.

Yipp, M.W. 1991. The relationship between hydrological factors and the distribution of freshwater gastropods in Hong Kong. *Verhandlungen Internationale Vereinigung für theoretische und angewandte Limnologie* 24: 2954–9.

Yipp, M.W., Cha, M.W. and Liang, X.Y. 1992. A preliminary impact assessment of the introduction of two species of *Ampullaria* (Gastropoda: Ampullariidae) into Hong Kong. In *Proceedings of the Tenth International Malacological Congress, Tübingen 1989*. (C. Meier-Brook, ed.), 393–7. Baja, Hungary: UNITAS Malacologia.

Zhao, E.M. and Adler, K. 1993. *Herpetology of China*. Oxford, Ohio: Society for the Study of Amphibians and Reptiles.

Index

HONG KONG CLIMATE

TEMPERATURE & RELATIVE HUMIDITY

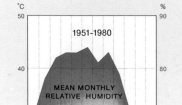
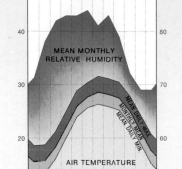

°C %

1951-1980

MEAN MONTHLY
RELATIVE HUMIDITY

MEAN DAILY MAX
MONTHLY MEAN
MEAN DAILY MIN

AIR TEMPERATURE

MONTH 1 2 3 4 5 6 7 8 9 10 11 12 MONTH

MONTHLY RAINFALL

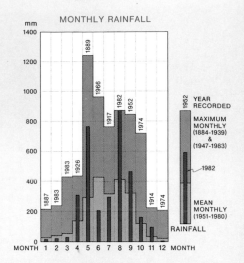

mm

YEAR RECORDED

MAXIMUM MONTHLY
(1884-1939)
&
(1947-1983)

1982

MEAN MONTHLY
(1951-1980)

RAINFALL

MONTH 1 2 3 4 5 6 7 8 9 10 11 12 MONTH

MEAN ANNUAL RAINFALL
1953-1982

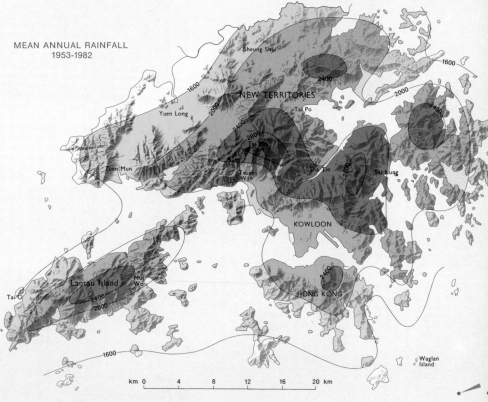

Sheung Shui

NEW TERRITORIES

Yuen Long

Tai Po

Tuen Mun

Tsuen Wan

Tai Mo Shan

Sha Tin

Sai Kung

KOWLOON

Lantau Island

Mui Wo

Tai O

HONG KONG

Waglan Island

km 0   4   8   12   16   20 km

Series AR/8/CM
Edition 1 1984

Cartography by Survey Division, Lands Department.
©Hong Kong Government.

knots 12.8   12.4

JAN   FEB